ELECTROTECHNICAL

LEARNER TRAINING MANUAL

AF685998

Name

City & Guilds enrolment number

Centre name

Centre number

Trainer

Date registered with City & Guilds

ACKNOWLEDGEMENTS

ABOUT THE AUTHOR: JOE CIESZYNSKI

Originally qualifying as an electronics engineer, Joe has worked as a service engineer, a full-time lecturer teaching electronic and electrical engineering, and as a freelance consultant developing distance learning materials for the electronic and electrical industries. Since 2003 Joe has worked for a large international electronic security company where he works in hardware/software development, training engineers on the installation, and in the commissioning and maintenance of the latest electronic security equipment. As well as classroom-based training, he also develops short training videos specifically to support engineers when working on site. He has published several textbooks.

SPECIAL THANKS

To Andy Jeffery for his content review.

ILLUSTRATONS, PHOTOGRAPHS AND FORMS

Andrew Hay-Ellis: page 14.
IET: pages 80–83, 94–98 and 108.
William Padden: pages 9, 10, 12, 15, 16, 17, 18, 19, 21, 23, 25, 27, 29, 30, 38, 39, 41, 43, 47, 49, 51, 52, 58, 63, 66, 68, 69, 72, 74, 75, 89, 90, 100, 106, 110, 112, 113, 115, 117, 118, 119, 121, 122, 123, 125 and 127.
Ann Paganuzzi: pages 13 and 111.
Screwfix: page 9.
Jules Selmes and Adam Giles (photographer and assistant): page 45.

CONTENTS

WELCOME TO THE CITY & GUILDS LEARNER TRAINING MANUAL

USING THIS MANUAL

Welcome to the City & Guilds Learner Training Manual.

The tasks in this manual are provided to support electrotechnical learners. There is no requirement to perform the tasks in the order given in this manual, and tutors should feel free to assign each task in relation to their scheme of work. This is especially true of tasks E40–E45, the science tasks.

In some cases the physical dimensions of the components or installation are given in the task, but in many cases the dimensions have been omitted because each centre will have different facilities, and the tutor/trainer will need to adapt the dimensions to suit. Some tasks require the learner to reference guidance documents such as the IET On-site Guide for dimensions, support distances and good installation practices.

Many of the activities are repeated over a number of tasks in order to offer learners the opportunity to develop their skills. Similarly, where a learner has not performed well in the execution of a task, the task may be repeated. Although many of the tasks are standalone, some of them (notably tasks 25–38) are designed to integrate into one combined task, providing learners with an opportunity to work with more complex installations. This also reduces material costs (a factor which cannot be ignored).

These tasks are not assessments and therefore are not compulsory.

PROGRESS TRACKING

Task number	Training task	Date completed	Tutor/trainer's initials
E1	Cable preparation		
E2	Conduit cutting		
E3	13A flexible cable connections		
E4	Steel trunking		
E5	Protective bonding		
E6	Bend and join steel conduit		
E7	Bend and join plastic conduit		
E8	Component alignment		
E9	Cable clipping and socket outlet connection		
E10	Safe isolation and locking-off: single-phase circuit		
E11	Installing PVC conduit and single-core cables		
E12	Installing steel conduit and single-core cables		
E13	One-way lighting circuit		
E14	Two-way lighting circuit		
E15	Two-way and intermediate lighting circuit		
E16	Testing: two-way and intermediate lighting circuit		
E17	Cable tray fabrication: flat 90° bend, joint and 45° set		
E18	Cable tray fabrication: tray reduction		
E19	SWA cable installation		
E20	MICC termination		
E21	FP200 termination		
E22	PVC trunking fabrication		
E23	Steel trunking fabrication		
E24	Planning an installation		
E25	System installation: ring final circuit		
E26	Safe isolation and locking-off: single-phase		
E27	System installation: two-way lighting circuit		
E28	System installation: steel trunking installation		
E29	System installation: radial circuit in steel conduit		
E30	System installation: protective bonding		
E31	System installation: transformer and extra-low-voltage lighting		
E32	System installation: incoming supply (simulated)		

Task number	Training task	Date completed	Tutor/trainer's initials
E33	Initial verification testing: single-phase system		
E34	Test results analysis		
E35	Three-phase distribution board		
E36	Periodic inspection and testing: three-phase system		
E37	Three-phase motor installation and testing		
E38	Fault location		
E39	Data cable and socket connection		
E40	Solder connections		
E41	Introduction to test instruments		
E42	Resistive circuit theory		
E43	Applied resistive circuit theory		
E44	Three-phase supply measurement		
E45	Component identification		

E1

CABLE PREPARATION

Note: your tutor/trainer will issue you with suitable lengths of cable.

Item	Complete all the tasks below	Candidate check ✓
1	Identify each cable type by writing its name in the space provided.	☐
2	Using an appropriate tool, remove 40mm of the outer sheath of cables A, B and C.	☐
3	Using an appropriate tool, remove the insulation on each conductor to reveal 10mm of bare conductor.	☐

For each cable, there should be no damage to the insulation, and no bare conductor showing as a result of such damage.

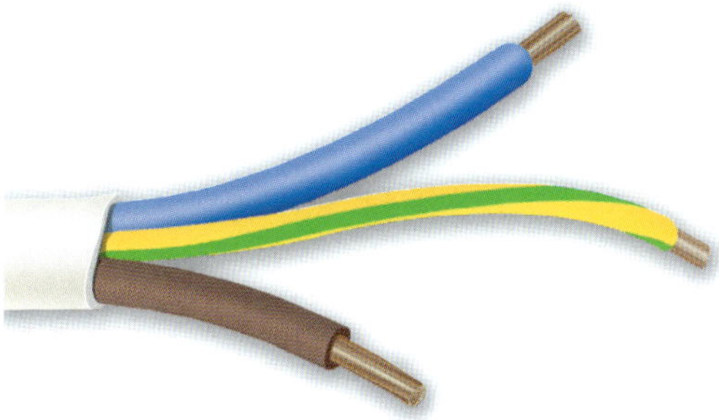

Cable type A: ______________________

Cable type B ______________________

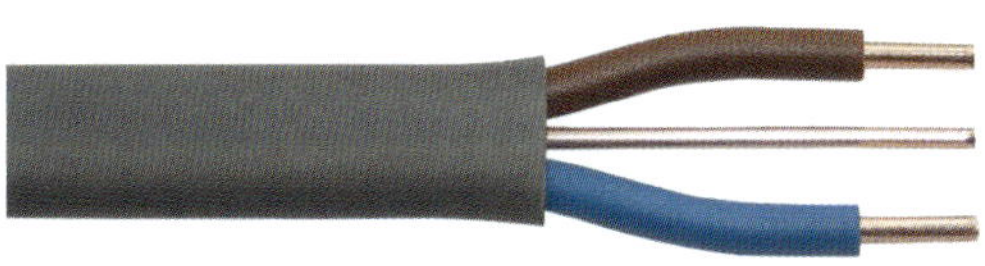

Cable type C: ______________________

Cable type D ______________________

Date completed

Tutor/trainer's feedback/comments

E2

CONDUIT CUTTING

Your tutor/trainer will provide you with lengths of plastic and steel conduit.

Item	Complete all the tasks below	Candidate check ✓
1	Using conduit cutters and a junior hacksaw, cut two 50mm lengths of plastic conduit.	☐
2	Using a junior hacksaw, cut two 50mm lengths of steel conduit.	☐
3	Remove all burrs from the ends of the cut pieces.	☐

When completed, stand the four pieces up on a flat surface as illustrated below to check that the cuts are straight and all lengths are equal.

Date completed

Tutor/trainer's feedback/comments

E3

13A FLEXIBLE CABLE CONNECTIONS

Your tutor/trainer will provide you with the 230V connector types illustrated below, and a length of three-core flexible cable.

Note: the dimensions below are indicative and can be used in the absence of any manufacturer's instructions.

Item	Complete all the tasks below	Candidate check ✓
1	Remove the outer sheath at each end of the cable (suggested between 40mm–50mm). Check that the inner insulation has not been cut or damaged.	☐
2	At one end of the cable, measure the conductors against the 13A plug and cut them to the correct lengths.	☐
3	Remove 7mm of insulation from each of the conductors intended for the 13A plug, and twist the copper strands.	☐
4	Insert the conductors into the correct terminals, making certain that they are secure and that there is no bare conductor showing at each terminal, nor any stray strands.	☐
5	Check the correct fuse rating for the cable being used (use your notes or other sources of information).	☐
6	Fit the cable restraint and the plug top.	☐
7	At the other end of the cable, measure the conductors against the IEC connector and cut them to the correct lengths.	☐
8	Remove 10mm of insulation from each of the conductors and twist the copper strands.	☐
9	Insert the conductors into the correct terminals, making certain that they are secure and that there is no bare conductor showing at the terminal, nor any stray strands.	☐
10	Fit the cable support and restraint and the plug top.	☐
11	Check the continuity and polarity of the lead using a low resistance ohmmeter or other suitable test instrument.	☐

Continued

E3 (CONTINUED)

13A FLEXIBLE CABLE CONNECTIONS

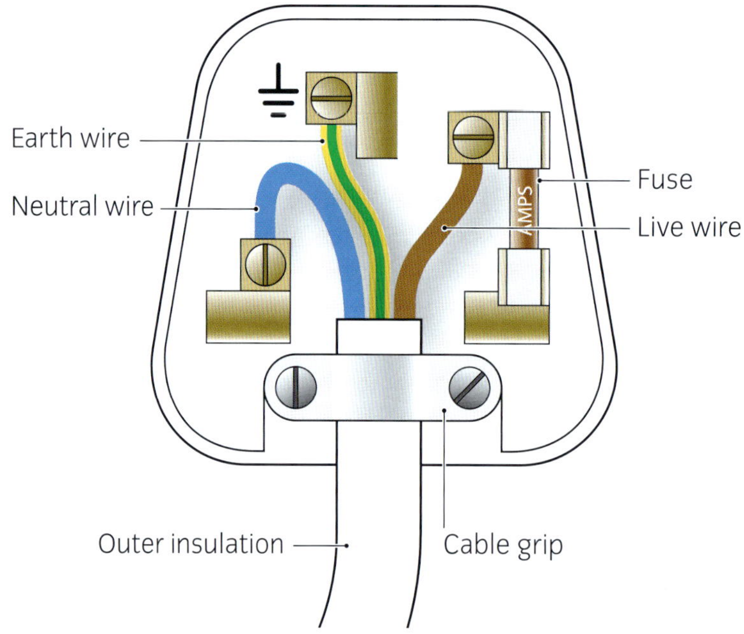

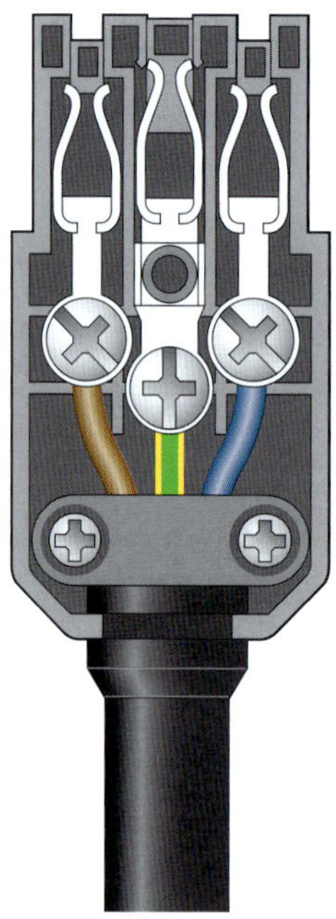

Date completed
Tutor/trainer's feedback/comments

E4

STEEL TRUNKING

Your tutor/trainer will provide you with a length of 50mm x 50mm steel trunking.

Item	Complete all the tasks below	Candidate check ✓
1	Using an engineer's square, measure and mark out a length of 95mm.	☐
2	Pack the trunking with wood in preparation for cutting.	☐
3	Cut to the required length using a hacksaw.	☐
4	Remove all burrs.	☐
5	Mark out a centre for a hole in either side of the trunking. Both holes must be in the dead centre of each side.	☐
6	Drill a 20mm hole in each side of the trunking.	☐
7	Remove all burrs from the two holes.	☐

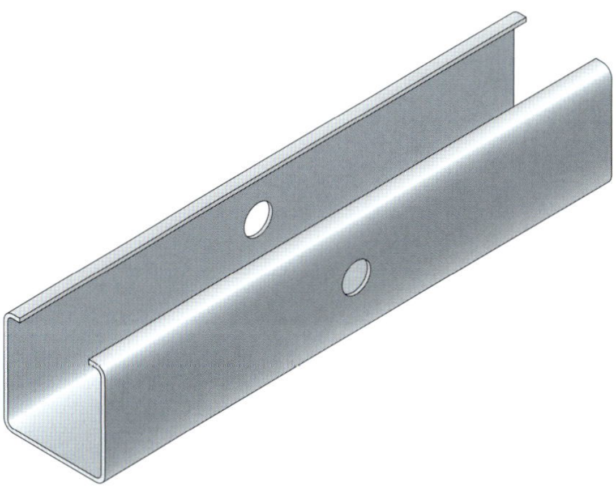

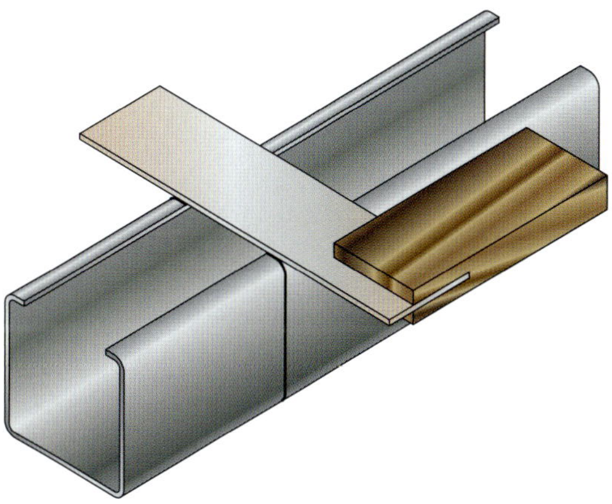

Date completed

Tutor/trainer's feedback/comments

E5

PROTECTIVE BONDING

Your tutor/trainer will provide you with a length of copper pipe, two pipe clips, a length of earth conductor and the components required to form an earth bond to the pipe.

Item	Complete all the tasks below	Candidate check ✓
1	Fix the pipe to a surface using the two clips.	☐
2	Clean the copper pipe in the area where the clamp is to be fitted.	☐
3	Wrap the clamp strap around the pipe and tighten until it is secure.	☐
4	Remove the insulation from the protective bonding conductor, sufficient to fit into the lug without excess bare copper being exposed.	☐
5	Using a crimping tool, secure the lug to the protective bonding conductor, then secure to the clamp.	☐
6	Ensure that the label is securely fitted.	☐
7	Dismantle the assembly once your tutor/trainer has confirmed it is satisfactory.	☐

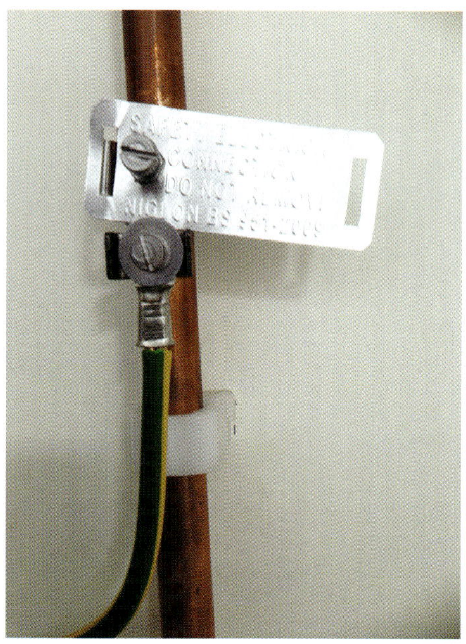

Date completed

Tutor/trainer's feedback/comments

E6

BEND AND JOIN STEEL CONDUIT

Your tutor/trainer will provide you with a length of 20mm steel conduit and a coupler.

Item	Complete all the tasks below	Candidate check ✓
1	Cut two lengths of conduit: 200mm and 500mm.	☐
2	Ream and de-burr ends as required.	☐
3	Thread both ends of each piece of conduit.	☐
4	Bend the 500mm conduit to 90° at the dimension shown below.	☐
5	Join the pieces using the coupler.	☐
6	Unscrew conduit from coupler after inspection by tutor/trainer.	☐

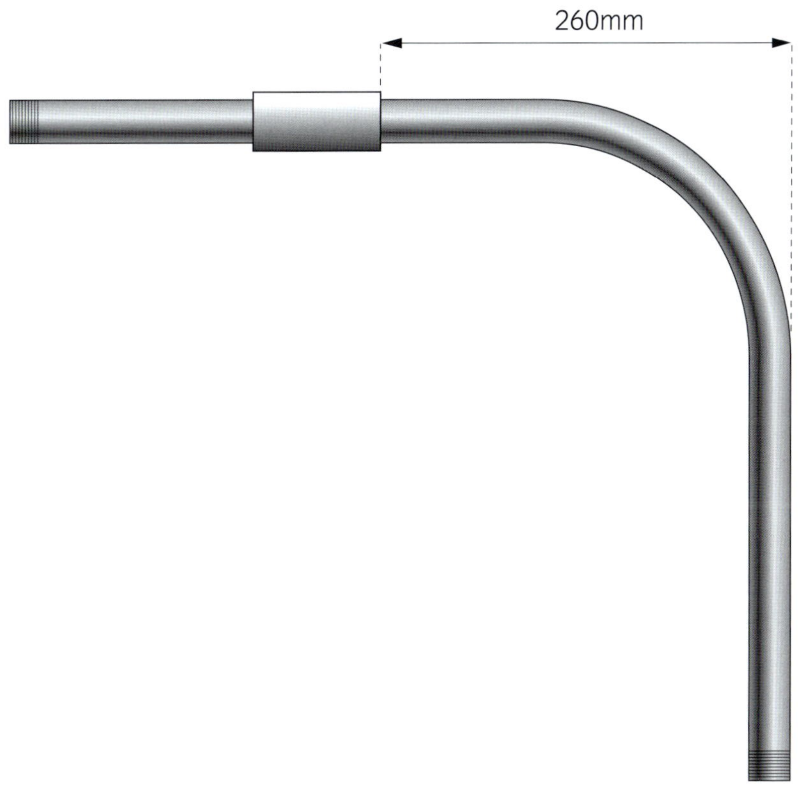

Date completed

Tutor/trainer's feedback/comments

E7

BEND AND JOIN PLASTIC CONDUIT

Your tutor/trainer will provide you with a length of 20mm plastic conduit and a coupler.

Note: this work may be used in task E11.

Item	Complete all the tasks below	Candidate check ✓
1	Cut two lengths of conduit: 600mm and 400mm.	☐
2	Ream and de-burr ends as required.	☐
3	Bend the 600mm conduit to 90° at the dimension shown below.	☐
4	Bend the 400mm conduit to form the offset shown below.	☐
5	Fix the conduit to the required surface, using the coupler to join the pieces.	☐

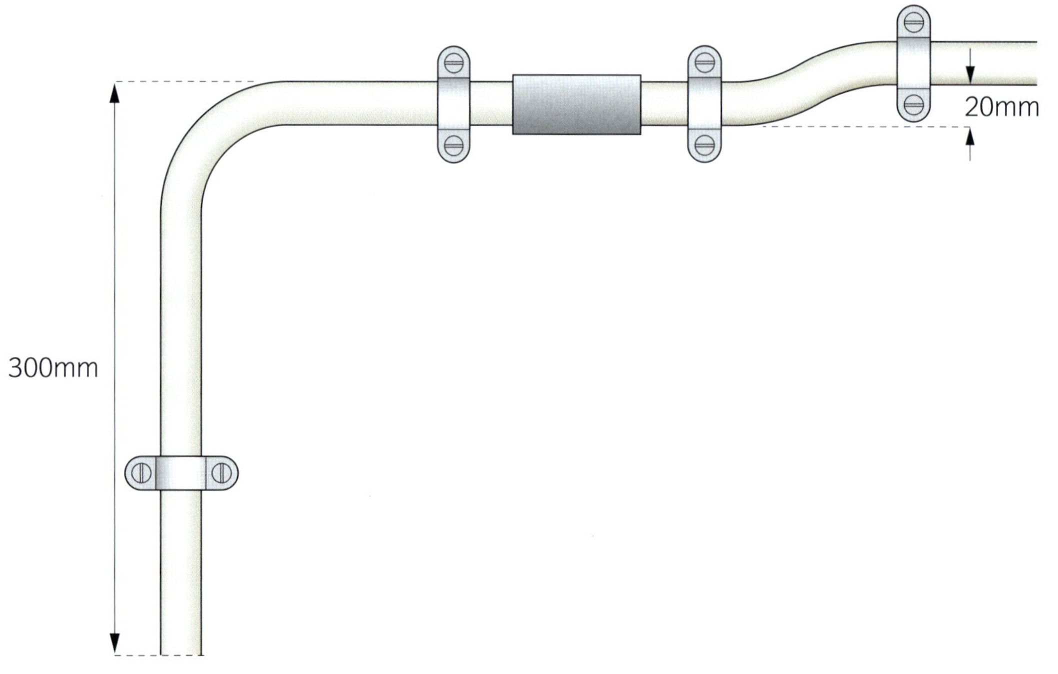

Date completed

Tutor/trainer's feedback/comments

E8

COMPONENT ALIGNMENT

Your tutor/trainer will provide you with a consumer unit, three surface metal back boxes, three socket-outlets and a range of fixings.

Note: this work will be used in task E9.

Item	Complete all the tasks below	Candidate check ✓
1	Measure, level and mark out the components on the surface.	☐
2	Select suitable fixings for the surface.	☐
3	Fix the components to the surface.	☐
4	Cut out either side or rear entry holes in the consumer unit suitable (your tutor/trainer will advise) for 2.5mm^2 flat profile twin and cpc cable entry points.	☐
5	When completed, check measurements to confirm correct installation.	☐

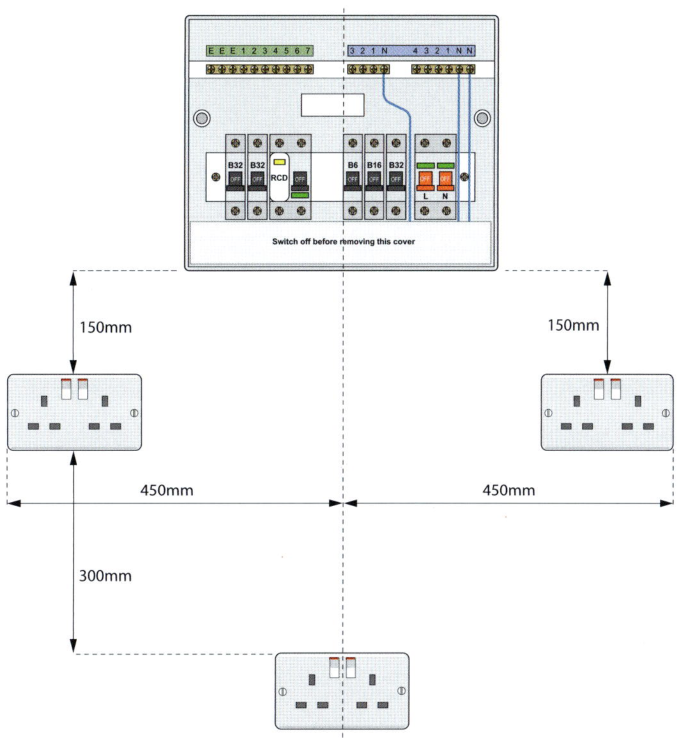

Date completed

Tutor/trainer's feedback/comments

E9
CABLE CLIPPING AND SOCKET OUTLET CONNECTION

Construct a final ring circuit using 2.5mm2 flat profile twin and cpc cable clipped to a surface.

Note: use the installation from task E8. This work will be used in task E10.

Item	Complete all the tasks below	Candidate check ✓
1	Remove knock-outs from metal boxes, where required, and insert rubber grommets.	☐
2	Measure and cut the cables to the required lengths. Allow sufficient length at each end for comfortable connection.	☐
3	Clip cables to surface as per the layout diagram below, paying particular attention to the radius of the bends. The IET On-site Guide may be consulted for bend radius dimensions. Ensure that the cables lie flat on the surface at all points.	☐
4	Remove a suitable length of outer sheath on each of the cable ends.	☐
5	Remove the insulation from each conductor, sufficient to fit into the terminals without excess bare copper being exposed.	☐
6	Fit green/yellow sleeve over all cpc conductors.	☐
7	Correctly connect the circuit to form a ring final circuit, as shown in the connection diagram.	☐
8	Ensure that all terminations are correct, and that they are electrically and mechanically sound.	☐

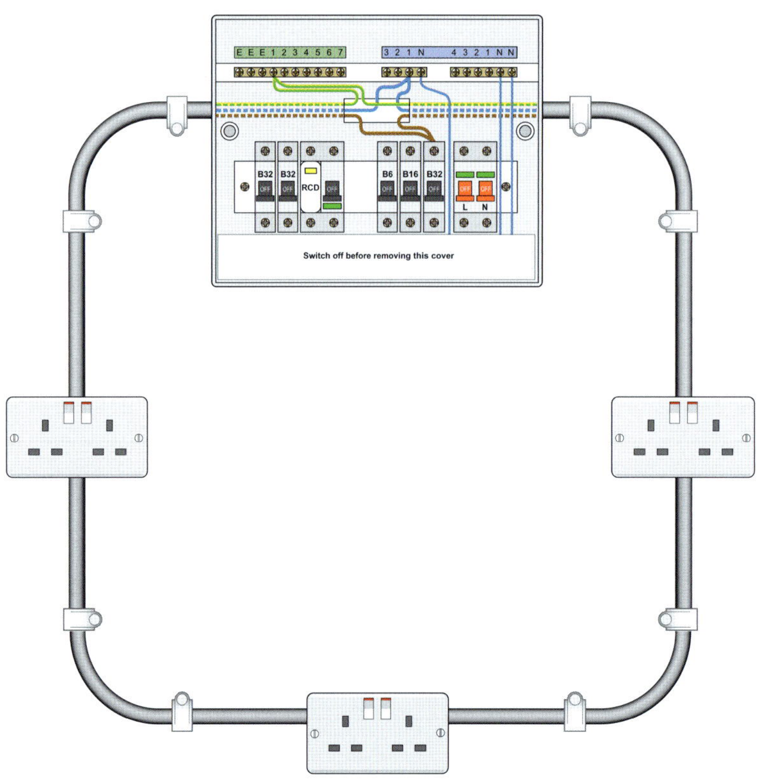

Continued

E9 (CONTINUED)

CABLE CLIPPING AND SOCKET OUTLET CONNECTION

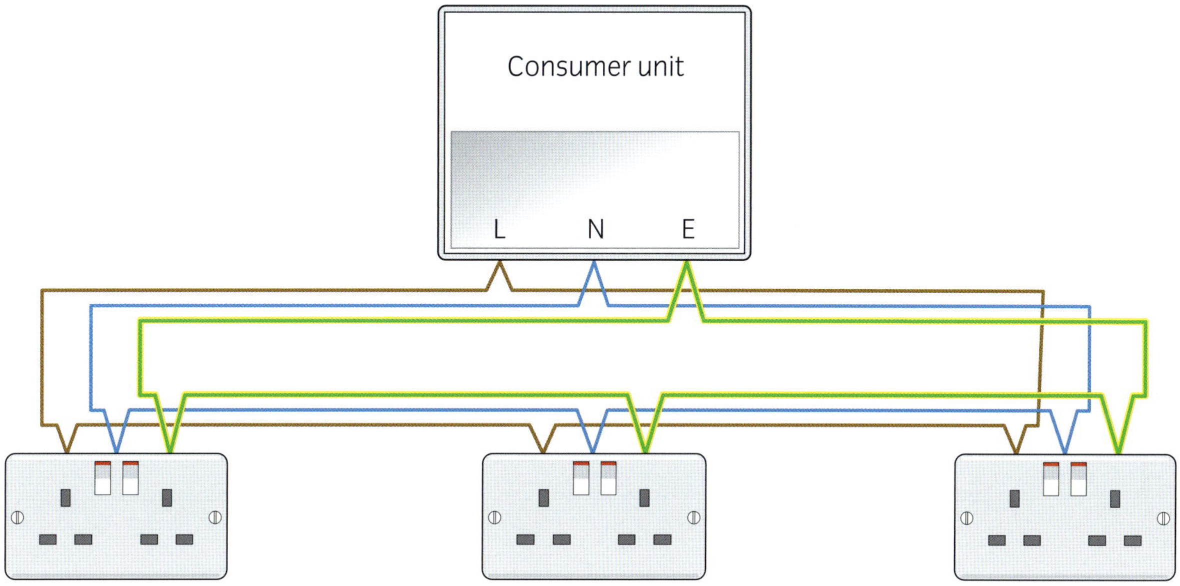

Date completed
Tutor/trainer's feedback/comments

E10

SAFE ISOLATION AND LOCKING-OFF: SINGLE-PHASE CIRCUIT

Assume that you are replacing one of the socket-outlets in the installation from task E9. You would first need to ensure that the circuit was safe to work on.

Note: use the installation from task E9.

The importance of safe isolation and locking-off procedure cannot be emphasised enough. Have you learned it yet? Prove it. Write the procedure here.

Continued

E10 (CONTINUED)

SAFE ISOLATION AND LOCKING-OFF: SINGLE-PHASE CIRCUIT

Item	Complete all the tasks below	Candidate check ✓
1	Ask your tutor/trainer to check your written procedure before proceeding.	☐
2	Isolate the circuit and use an appropriate locking-off kit to secure it.	☐
3	Following the correct method, use an appropriate test instrument to prove that the circuit is dead and safe to work on.	☐
4	Reinstate the supply.	☐
5	Dismantle the assembly once your tutor/trainer has confirmed this task is completed to satisfaction.	☐

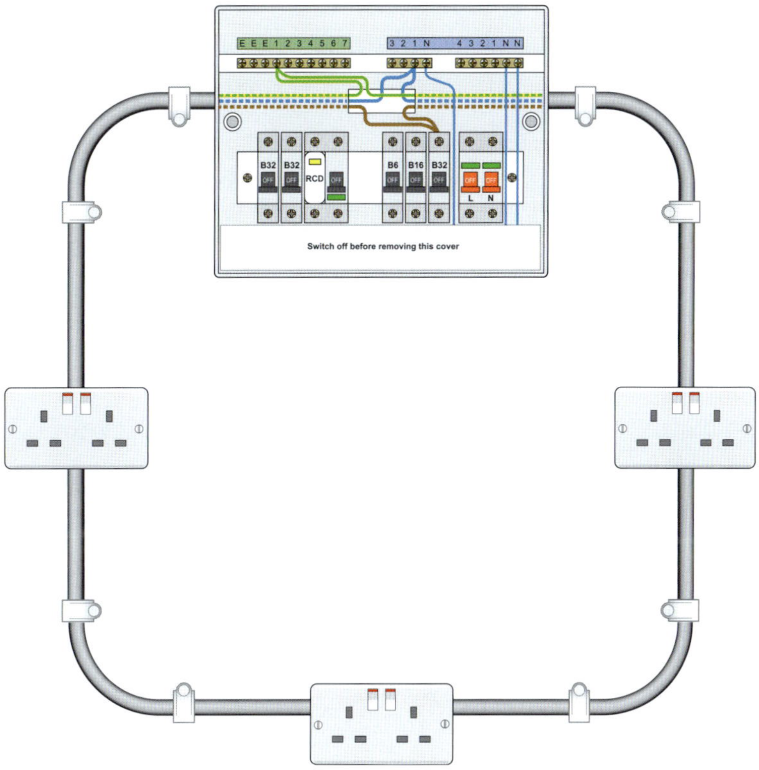

Date completed

Tutor/trainer's feedback/comments

E11

INSTALLING PVC CONDUIT AND SINGLE-CORE CABLES

Connect two 13A socket-outlets using 2.5mm^2 single cables in PVC conduit.

Note: you will adapt the conduit assembly from task E7.

Item	Complete all the tasks below	Candidate check ✓
1	Fix the two socket back boxes to a surface, using the dimensions given below.	☐
2	Using the conduit assembly from task E7, measure and cut the vertical length. Ream and de-burr as required.	☐
3	Connect the conduit to socket 1 back box, and secure using saddles.	☐
4	Measure and cut a piece of conduit to extend the conduit to socket 2. Ream and de-burr as required.	☐
5	Fix the offset conduit piece to the surface, and secure using a saddle.	☐
6	Measure and cut the line, neutral and cpc cables to the required lengths. Allow sufficient length at each end for comfortable connection.	☐
7	Pass the cables through the conduit.	☐
8	Complete the conduit installation using appropriate connectors.	☐
9	Remove the insulation from each conductor, sufficient to fit into the terminals without excess bare copper being exposed.	☐
10	Connect the line, neutral and cpc at each socket-outlet, making sure that terminations are electrically and mechanically sound.	☐
11	Secure the socket-outlets to the back boxes.	☐

Continued

E11 (CONTINUED)

INSTALLING PVC CONDUIT AND SINGLE-CORE CABLES

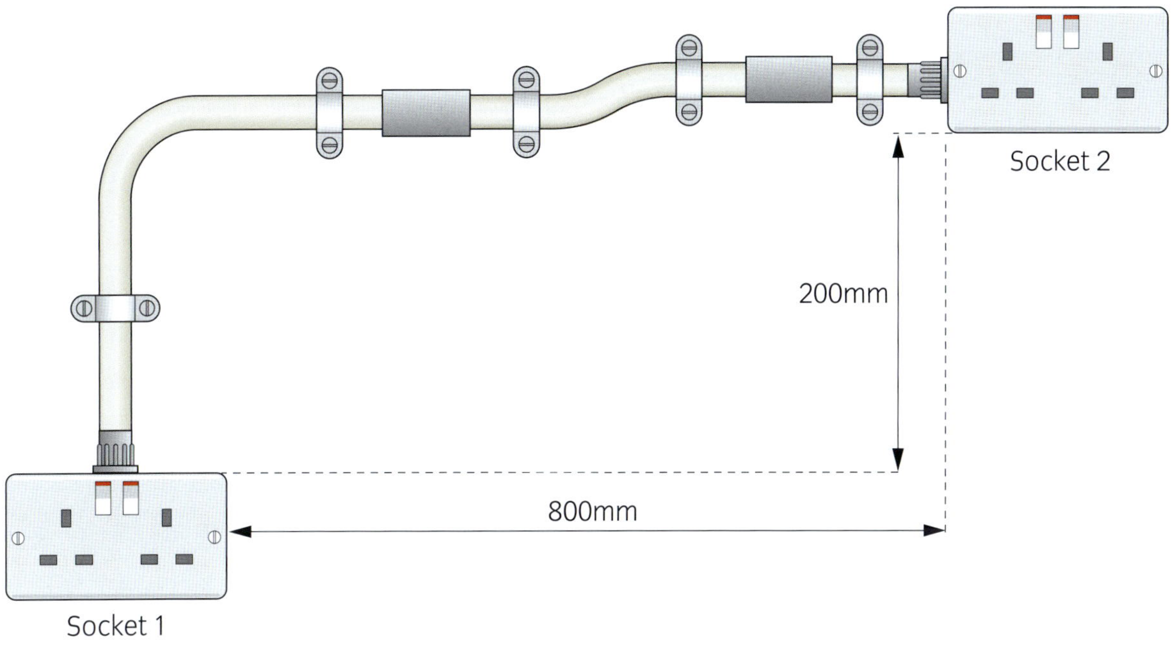

Date completed

Tutor/trainer's feedback/comments

E12

INSTALLING STEEL CONDUIT AND SINGLE-CORE CABLES

Your tutor/trainer will provide you with a length of 20mm steel conduit, a coupler, elbow, 13A socket-outlet, a metal back box and suitable fixings.

Item	Complete all the tasks below	Candidate check ✓
1	Fix the socket back box and elbow to a surface, using the measurements given in the diagram.	☐
2	Cut the conduit to the required lengths.	☐
3	Ream and de-burr as required.	☐
4	Thread all conduit ends.	☐
5	Bend the conduit to 90° using the measurements given.	☐
6	Install the conduit as shown in the diagram.	☐
7	Measure and cut the line, neutral and cpc cables to the required lengths. Allow sufficient length for comfortable connection.	☐
8	Pass the cables through the conduit and fit the elbow cover.	☐
9	Remove the insulation from each conductor, sufficient to fit into the terminals without excess bare copper being exposed.	☐
10	Connect the line, neutral and cpc at the socket outlet, making sure that terminations are electrically and mechanically sound.	☐
11	Dismantle the assembly once your tutor/trainer has confirmed it is satisfactory.	☐

Continued

E12 (CONTINUED)

INSTALLING STEEL CONDUIT AND SINGLE-CORE CABLES

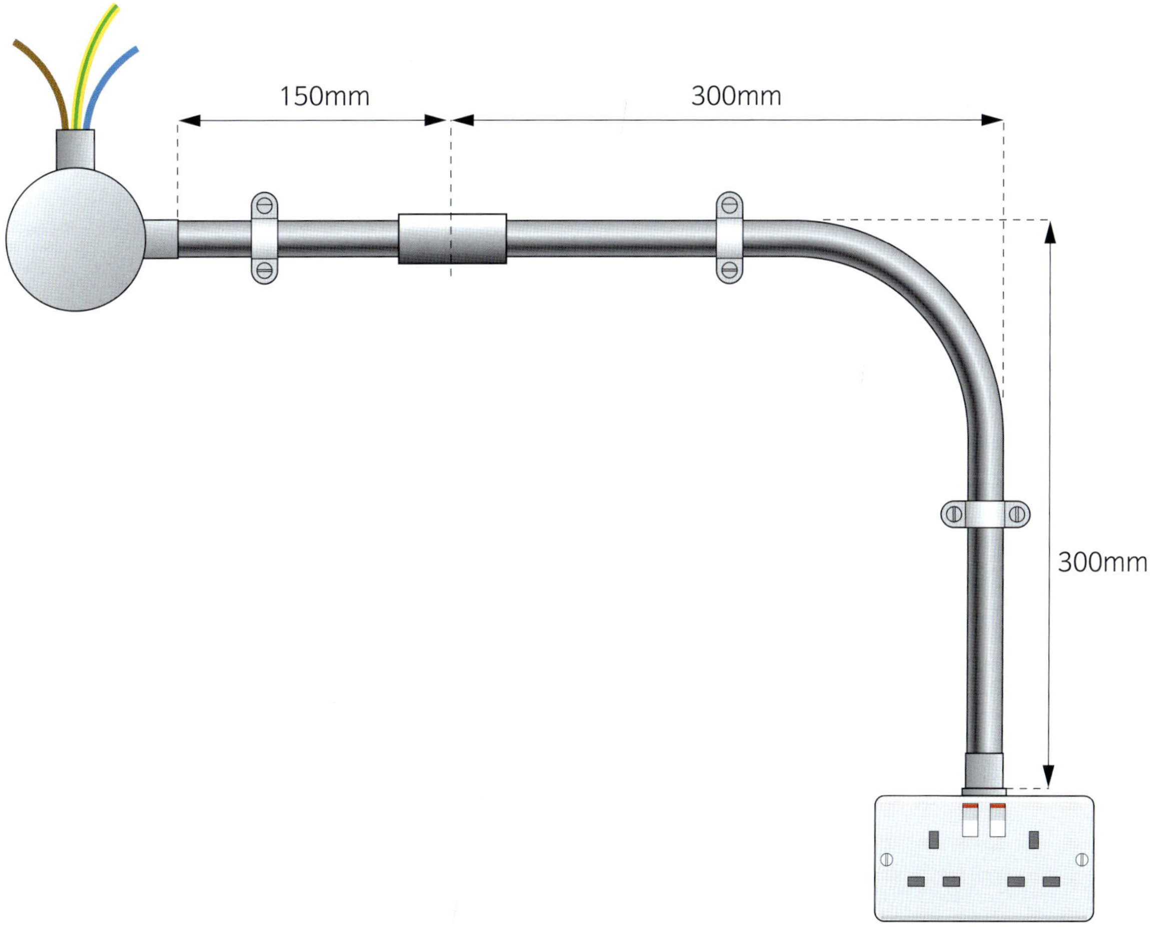

Date completed

Tutor/trainer's feedback/comments

E13

ONE-WAY LIGHTING CIRCUIT

Connect a single pendant light fitting and switch as a part of a lighting circuit, using 1.5mm^2 flat profile twin and cpc cable.

Note: your tutor may opt to provide you with a ready-made pendant fitting, or you may first have to assemble the lampholder and flexible cable before proceeding with the instructions below.

Item	Complete all the tasks below	Candidate check ✓
1	Fix the ceiling rose and switch to a surface. Clip cables as shown in the diagram below. Note: the cables may be fixed to the rear of a board to enable rear entry for the switch and the ceiling rose.	☐
2	Remove a suitable length of outer sheath on each of the cable ends.	☐
3	Remove the insulation from each conductor, sufficient to fit into the terminals without excess bare copper being exposed.	☐
4	Fit green/yellow sleeve over each cpc.	☐
5	Make sure that all line conductors are clearly identified using coloured sleeving, as necessary.	☐
6	Connect the feed and switch-line cables as shown in the diagram below, making certain that terminations are electrically and mechanically sound.	☐
7	Connect the flexible cable as shown, making certain that the cables pass through the restraint hooks in the ceiling rose.	☐
8	Replace the ceiling rose cover, making certain that there is no flex visible outside of the ceiling rose.	☐
9	Dismantle the installation and secure the terminal screws after inspection by your tutor/trainer.	☐

Continued

E13 (CONTINUED)

ONE-WAY LIGHTING CIRCUIT

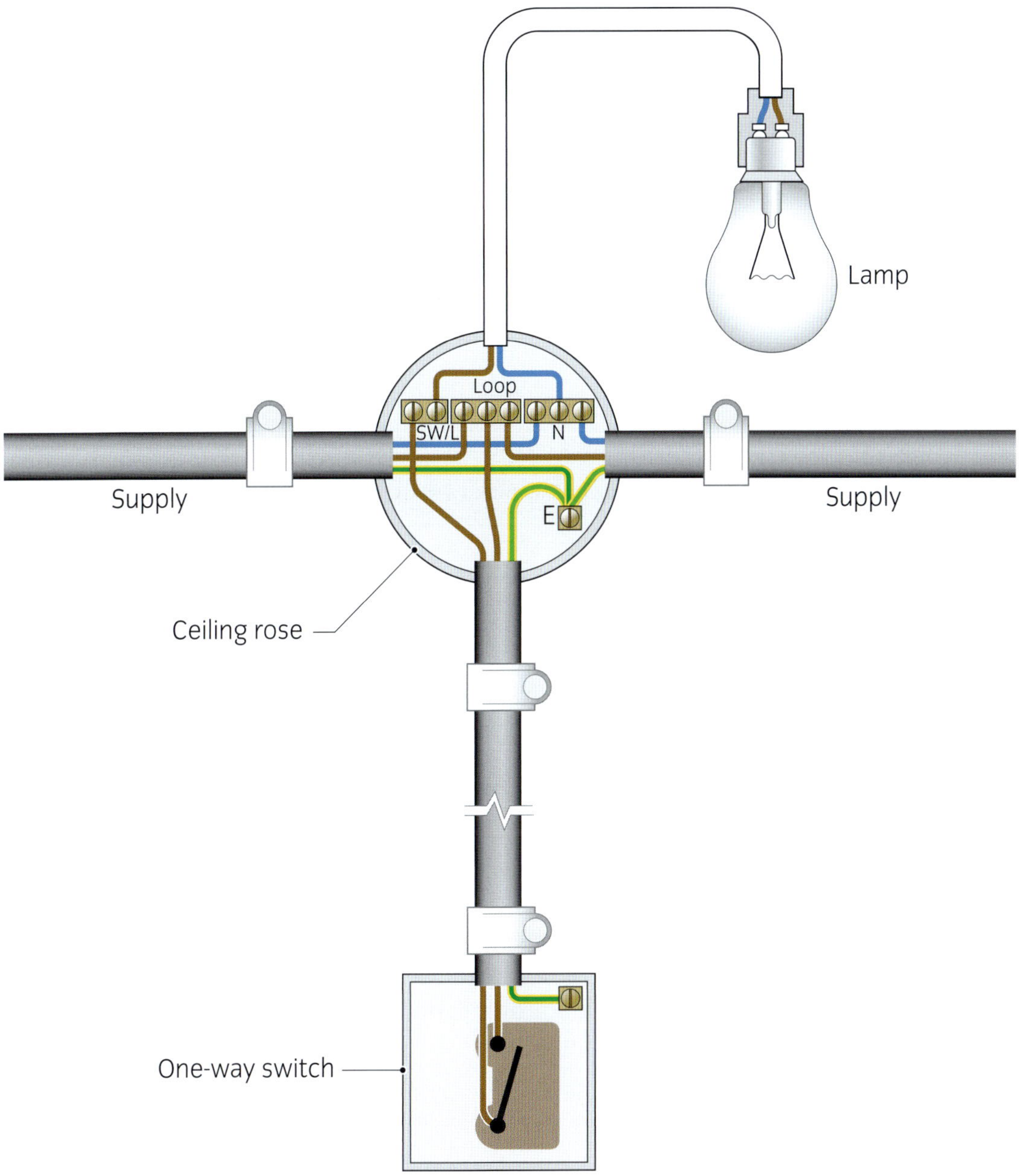

Date completed

Tutor/trainer's feedback/comments

E14

TWO-WAY LIGHTING CIRCUIT

Connect a single pendant light fitting and two-way switches as a part of a lighting circuit.

Note: your tutor will provide you with both two-core and cpc, and three-core and cpc flat profile cable, a single pendant fitting and two two-way switches.

Item	Complete all the tasks below	Candidate check ✓
1	Fix the ceiling rose and switches to a surface, and clip cables as shown in the diagram below. Note: the cables may be fixed to the rear of a board to enable rear entry for the switches and the ceiling rose.	☐
2	Remove a suitable length of outer sheath on each of the cable ends.	☐
3	Remove the insulation from each conductor, sufficient to fit into the terminals without excess bare copper being exposed.	☐
4	Fit green/yellow sleeve over each cpc conductor.	☐
5	Make sure that all line conductors are clearly identified using coloured sleeving, as necessary.	☐
6	Connect the feed and switch-line cables as shown in the circuit diagram, making certain that terminations are electrically and mechanically sound.	☐
7	Connect the flexible cable as shown in the circuit diagram, making certain that the cables pass through the restraint hooks in the ceiling fitting.	☐
8	Replace the ceiling rose cover, making certain that there is no flex visible outside of the ceiling rose.	☐
9	Check the circuit operation by connecting a low resistance ohmmeter between the 'loop' and 'L' terminals in the ceiling rose. Operating the switches in any order should give alternate 'open' and 'closed' circuit readings.	☐
10	Dismantle the installation and secure the terminal screws after inspection by your tutor/trainer.	☐

Continued

E14 (CONTINUED)

TWO-WAY LIGHTING CIRCUIT

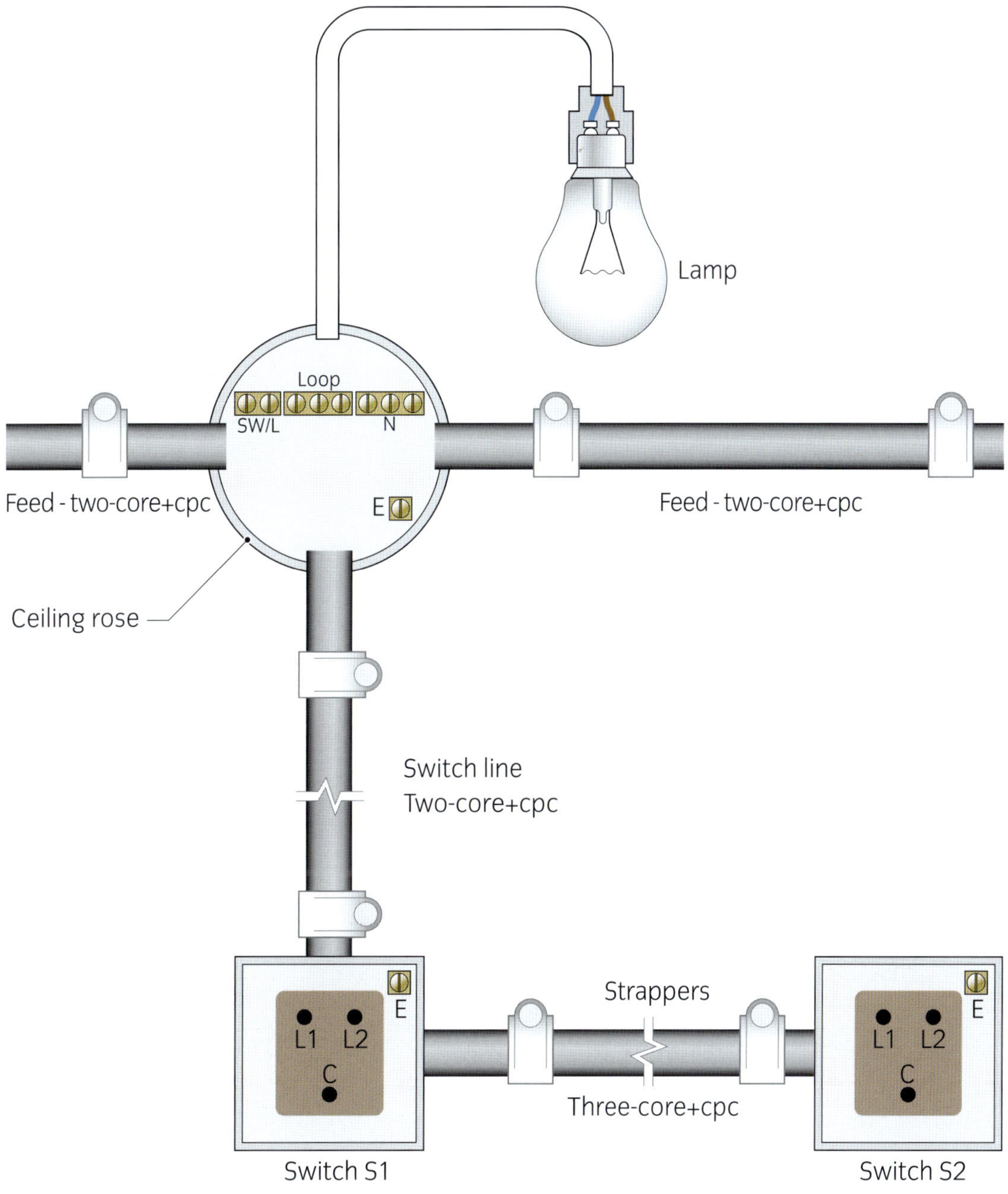

E14 (CONTINUED)

TWO-WAY LIGHTING CIRCUIT

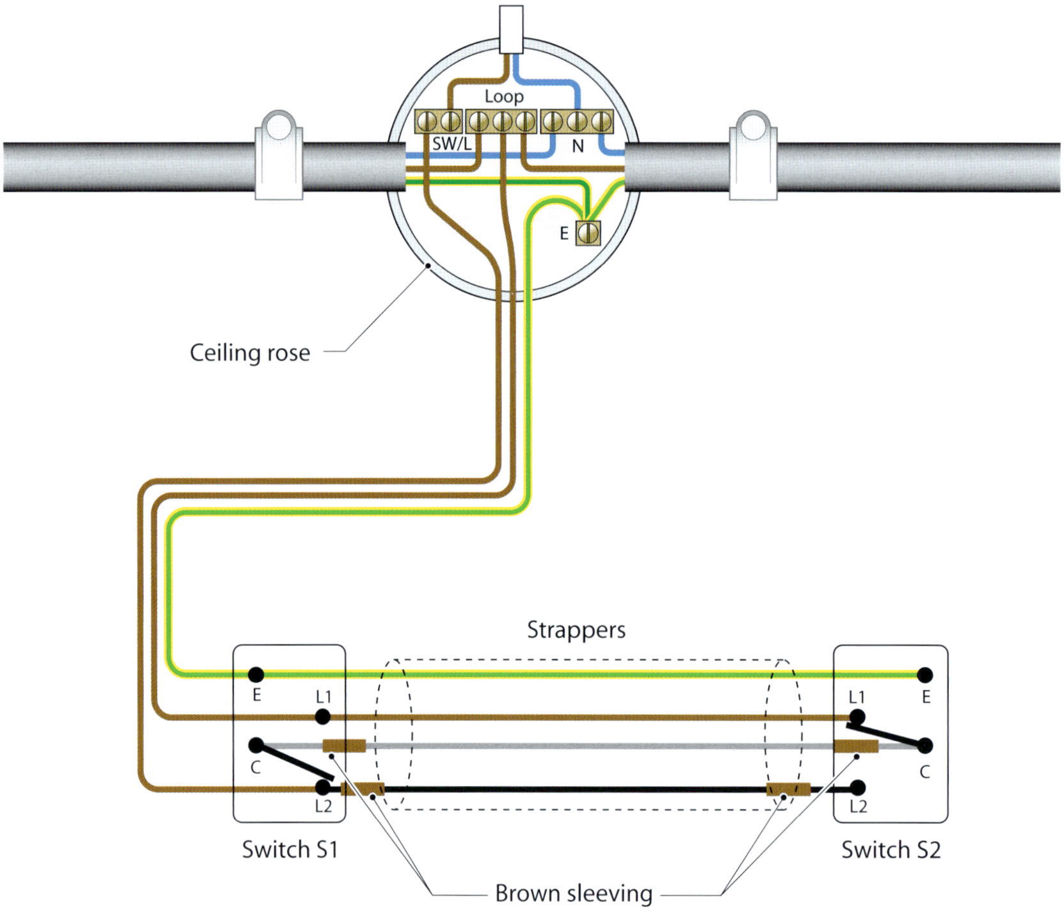

Date completed

Tutor/trainer's feedback/comments

E15

TWO-WAY AND INTERMEDIATE LIGHTING CIRCUIT

Connect a single pendant light fitting to a two-way switch circuit that includes a third intermediate switch, as part of a lighting circuit.

In a two-way lighting circuit with an intermediate switch, any switch will turn the light *on* or *off*. Such a circuit may be used on a stairway with multiple landings or in a long corridor. With this arrangement, many more intermediate switches could be added, allowing many switching points if required.

Using your notes or other sources of information, find a practical circuit diagram and copy it into your Learner Training Manual in the space below. Your circuit should include both a loop in and loop out section of the circuit.

Note: this work will be used in task E16.

Circuit diagram

Continued

E15 (CONTINUED)

TWO-WAY AND INTERMEDIATE LIGHTING CIRCUIT

Procurement list

Continued

E15 (CONTINUED)

TWO-WAY AND INTERMEDIATE LIGHTING CIRCUIT

Item	Complete all the tasks below	Candidate check ✓
1	Ask your tutor to check your circuit before proceeding.	☐
2	In the space provided, produce a procurement list for the components and materials that you will require for this task.	☐
3	Obtain the items in your list from your stores.	☐
4	Fix the components to a surface, and fix cables necessary to construct the circuit in your diagram. Note: the cables may be fixed to the rear of a board to enable rear entry for the switches and the ceiling rose.	☐
5	Remove a suitable length of outer sheath on each of the cable ends.	☐
6	Remove the insulation from each conductor, sufficient to fit into the terminals without excess bare copper being exposed.	☐
7	Fit green/yellow sleeve over each cpc conductor.	☐
8	Make sure that all line conductors are clearly identified using coloured sleeve, as necessary.	☐
9	Connect the feed and switch-line cables, making certain that terminations are electrically and mechanically sound.	☐
10	Connect the flexible cable as shown, making certain that the cables pass through the restraint hooks in the ceiling fitting.	☐
11	Replace the ceiling rose cover, making certain that there is no flex visible outside of the ceiling rose.	☐
12	Check the circuit operation by connecting a low resistance ohmmeter between the 'loop' and 'SW/L' terminals in the ceiling rose. Operating the switches in any order should give alternate 'open' and 'closed' circuit readings.	☐

Date completed

Tutor/trainer's feedback/comments

E16

TESTING: TWO-WAY AND INTERMEDIATE LIGHTING CIRCUIT

Test the circuit functionality using an appropriate low reading ohmmeter.

Carry out insulation resistance testing on part of a lighting circuit.

Note: use the installation from task E15.

TEST 1 – CIRCUIT FUNCTIONALITY

Item	Complete all the tasks below	Candidate check ✓
1	Connect a known good lamp or LED equivalent to the pendant (not a fluorescent equivalent).	☐
2	Set your continuity meter to a resistance range in the order of 200Ω. This is necessary to make sure that it will read the lamp resistance, which is typically 40Ω–80Ω.	☐
3	Connect the test instrument to the loop and neutral terminals in the ceiling rose.	☐
4	Operate each switch in turn. You should see the test instrument switch between a reading and infinity as the lamp is connected and disconnected by the switch action.	☐
5	If your circuit is not operating correctly, carefully re-check all of your connections, making sure that they follow the circuit diagram, and that they are secure. Then re-test your circuit.	☐
6	Once you have proven that your circuit is functioning correctly, remove the lamp.	☐

Continued

E16 (CONTINUED)

TESTING: TWO-WAY AND INTERMEDIATE LIGHTING CIRCUIT

TEST 2 – INSULATION RESISTANCE

Note: it is not possible for you to perform testing of a complete lighting circuit because the installation in task E15 only represents a part of such a circuit. However, you can still practise the principles of insulation resistance testing using this simple circuit.

Item	Complete all the tasks below	Candidate check ✓
1	During this test, the line, neutral and cpc conductors should be connected to the terminals of a consumer unit/DB. For the purpose of this exercise, we will assume that this is the case. Ensure that the cable cores of the incoming feed and on-going feed are separated.	☐
2	Set the test instrument to the 500 V test range. Remember that you are looking for a reading in excess of 1 MΩ (typical instrument reading will be >200 MΩ).	☐
3	Log the results of the following tests in the first column of the results table.	☐
4	In this task, in the absence of an actual consumer unit, you will have to perform your testing at the ceiling rose. Test the insulation resistance between: neutral and earth loop (line) and neutral loop (line) and earth	☐
5	Operate one of the switches (not the intermediate switch) and repeat the loop and earth test.	☐
6	Operate the intermediate switch and repeat the loop and earth test.	☐
7	Operate the third switch and repeat the loop and earth test.	☐

Continued

E16 (CONTINUED)

TESTING: TWO-WAY AND INTERMEDIATE LIGHTING CIRCUIT

Applied test	Test 2 results	Test 3 results
Neutral – earth		
Loop – neutral		
Loop – earth <1>		
Loop – earth <2>		
Loop – earth <3>		
Loop – earth <4>		

TEST 3 – FAULT SIMULATION

Now that you have seen how a working circuit appears under test, let's see what effect a faulty cable will have on your test results.

Item	Complete all the tasks below	Candidate check ✓
1	Connect a suitable 1kΩ resistor between any one of the live terminals on your intermediate switch and earth. This resistor represents a cable insulation fault between these two points.	☐
2	Repeat the tests outlined in steps 2–7 in test 2 – insulation resistance exercise.	☐

Date completed

Tutor/trainer's feedback/comments

E17

CABLE TRAY FABRICATION: FLAT 90° BEND, JOINT AND 45° SET

Construct a 90° bend from three pieces of light gauge perforated metal tray, and add an extension with a 45° taper, using a fishplate.

Agree the dimensions A, B and C in Figure 1 with your tutor/trainer and add them to the diagram before starting this task. Dimension C should be long enough to accommodate a tray reduction in task E18.

Assume that the tray assembly will be fixed to a vertical surface using 25mm stand-off brackets. (Check with your tutor/trainer on the size of bracket available in your stores).

Note: this work will be used in tasks E18–E19.

Item	Complete all the tasks below	Candidate check ✓
1	Measure and cut the four pieces of tray to the required lengths.	☐
2	Place tray pieces B and C at 90° to each other (Figure 2).	☐
3	Place the piece for the bend on top of these, and mark the points where each edge meets (Figure 3).	☐
4	Cut away the flanges on the two straight pieces between points a–b and a–c (Figure 4).	☐
5	Cut the diagonal piece along the lines g–e and f–d (Figure 5).	☐
6	Cut away the flanges between points c–g and b–f (Figure 5).	☐
7	De-burr all edges as necessary.	☐
8	Assemble the three pieces using suitable bolts/nuts.	☐
9	Take tray piece A and taper the end downwards by 45°. The cutting dimensions for this are shown in Figure 6. The taper should be set to match the 25mm stand-off brackets (or what ever size you have used for this task).	☐
10	Fit this piece to the bend using a fishplate and suitable bolts/ nuts (Figure 1).	☐

Continued

E17 (CONTINUED)

CABLE TRAY FABRICATION: FLAT 90° BEND, JOINT AND 45° SET

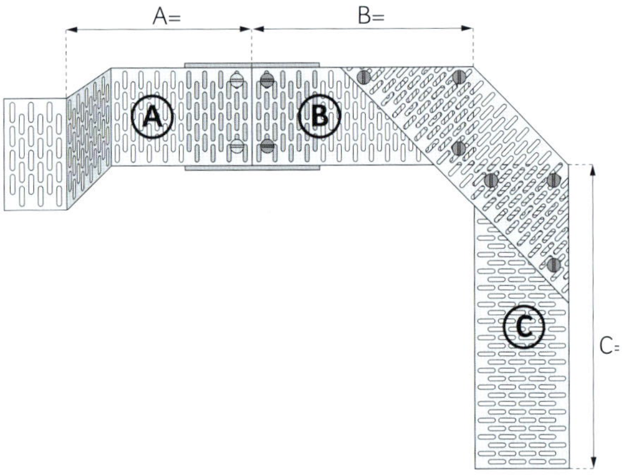

Figure 1

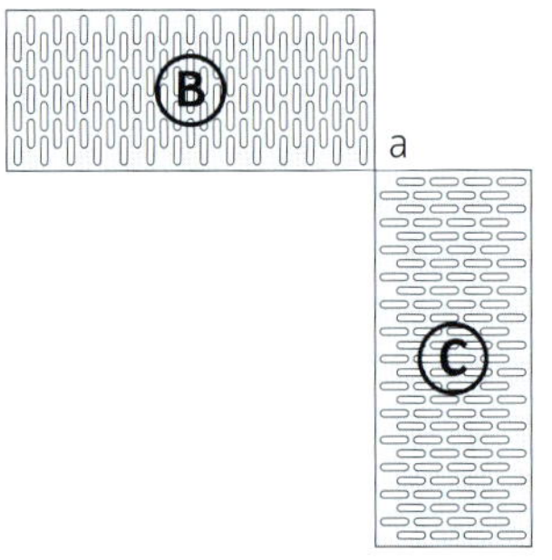

Figure 2

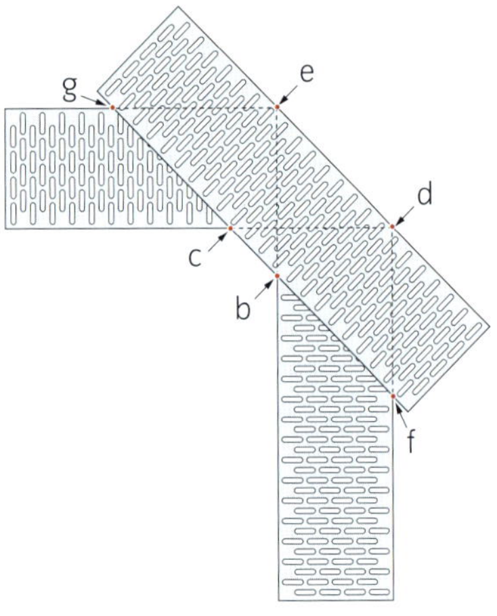

Figure 3

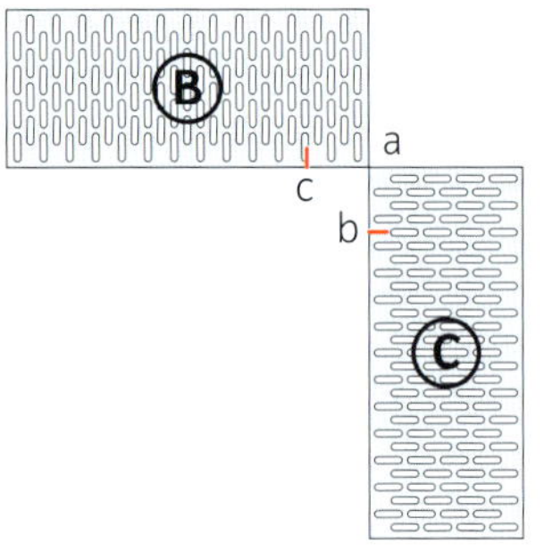

Figure 4

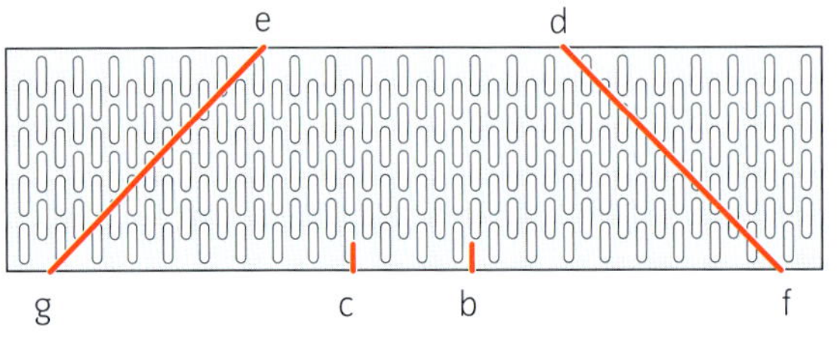

Figure 5

Continued

E17 (CONTINUED)

CABLE TRAY FABRICATION: FLAT 90° BEND, JOINT AND 45° SET

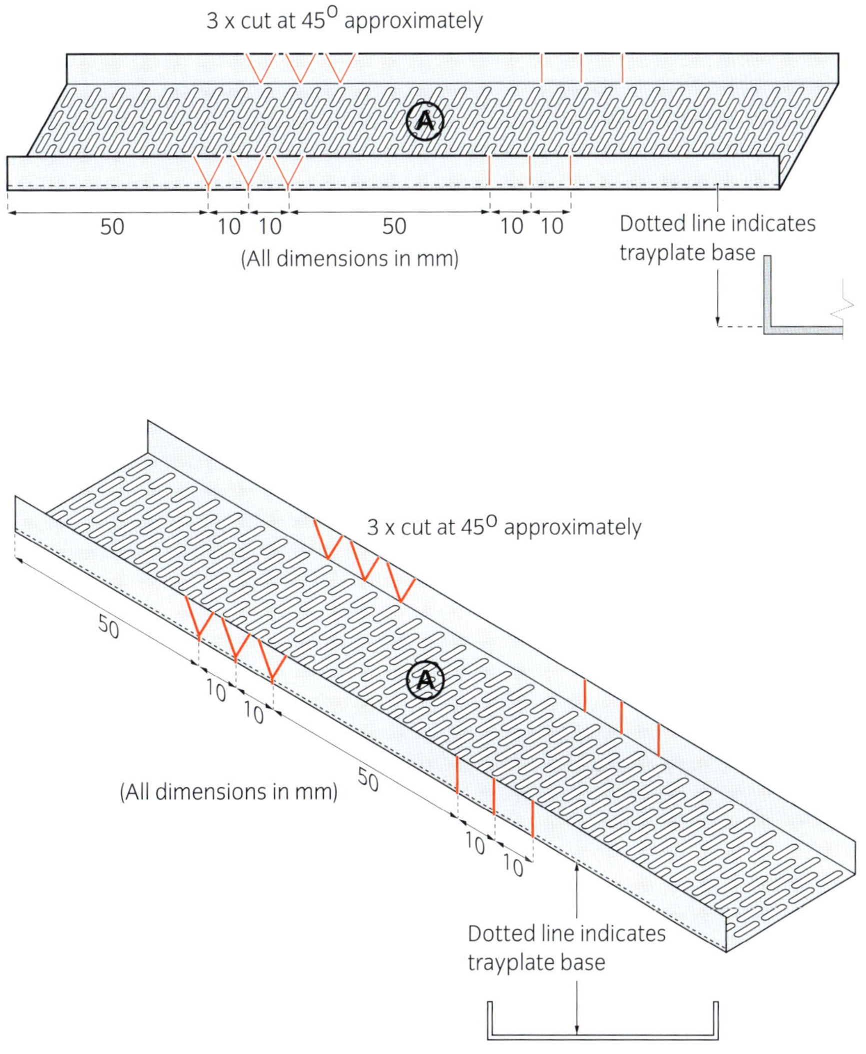

Figure 6

Date completed
Tutor/trainer's feedback/comments

E18

CABLE TRAY FABRICATION: TRAY REDUCTION

Construct a tray reduction and connect this to the item formed in task E17, using a fishplate.

Agree the dimensions for the tray reduction with your tutor/trainer and add them to the diagram before starting this task.

Note: use the tray assembled in task E17. This assembly will be used for task E19.

Item	Complete all the tasks below	Candidate check ✓
1	Dismantle Section C from the assembly in task E17, Figure 1. You will use this piece to create the tray reduction.	☐
2	Cut a piece of narrow tray to the length agreed with your tutor.	☐
3	Overlap the two pieces of tray by at least 100mm, and mark the required cutting line (Figure 1).	☐
4	Cut along the lines as shown in Figure 2, using a hacksaw and metal cutters. **Important:** Do not cut through the flanges.	☐
5	De-burr all edges as necessary.	☐
6	Lay the narrow piece back over the cut piece and fold into shape, passing the excess areas underneath the assembly (Figure 3).	☐
7	Fasten the sections together using suitable bolts/nuts.	☐
8	Re-connect the new piece to the assembly from task E17 (Figure 4).	☐

Continued

E18 (CONTINUED)

CABLE TRAY FABRICATION: TRAY REDUCTION

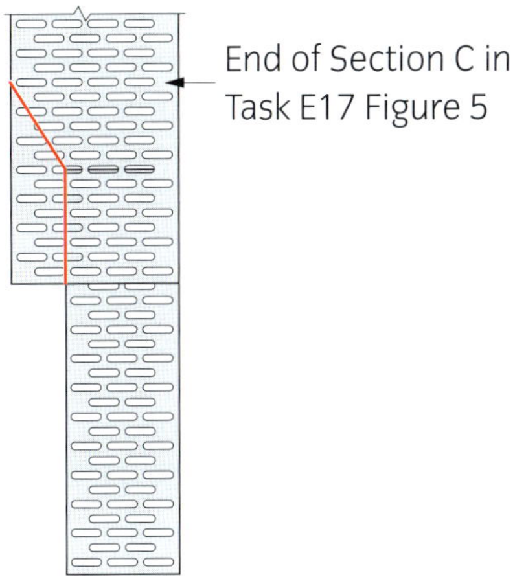

Figure 1

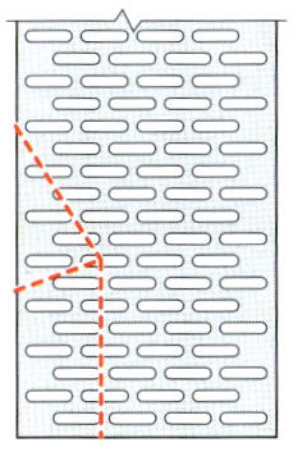

Figure 2

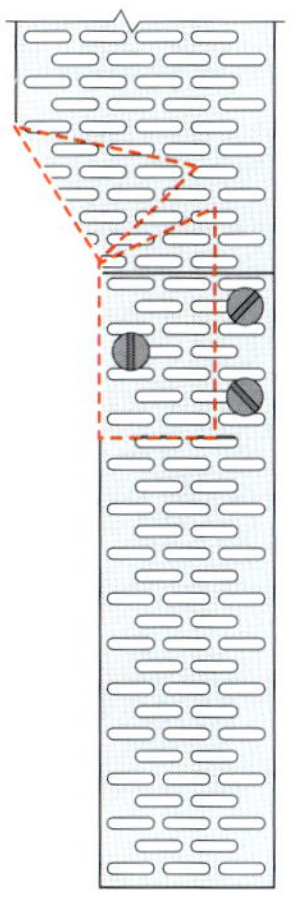

Figure 3

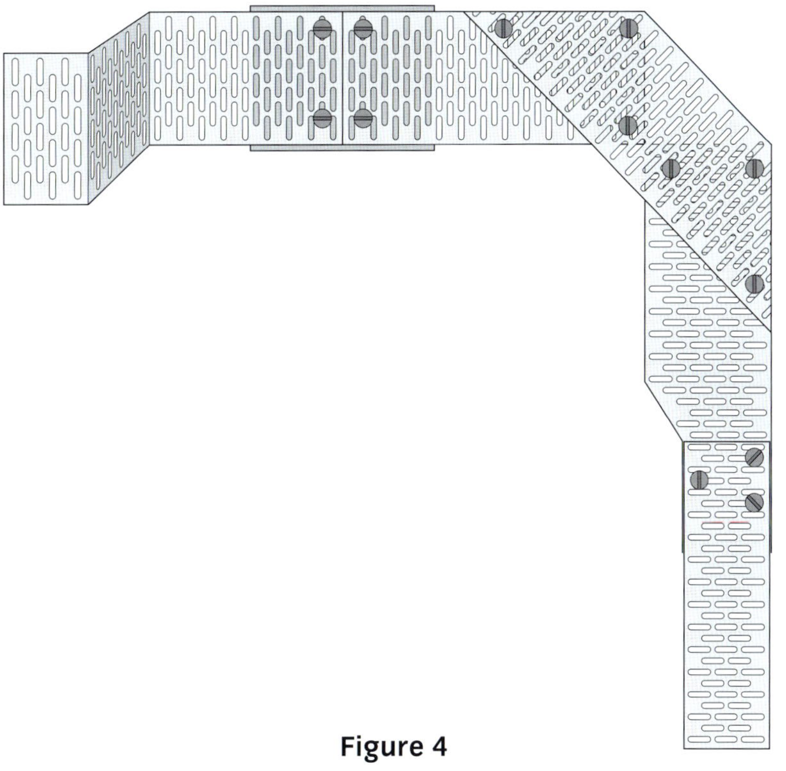

Figure 4

Date completed

Tutor/trainer's feedback/comments

E19

SWA CABLE INSTALLATION

Your tutor/trainer will provide you with a length of SWA two-core cable, a single-phase 230V commando socket, an adaptable metal box, a suitable gland and shroud, and cable ties.

Note 1: There are numerous instruction sets on SWA cable termination, including City & Guilds SmartScreen, and the City & Guilds textbook *Level 3 NVQ Diploma in Electrotechnical Technology*, Units 305–306. You may find it helpful to refer to these instructions during this task.

Note 2: If you are using 10mm SWA, you will use reducing lugs to secure the cable at the commando socket.

Item	Complete all the tasks below	Candidate check ✓
1	Fix the tray assembly from task E18 to a surface, using 25mm stand-off brackets.	☐
2	Fix the commando socket and adaptable box to the surface, about 150mm from the ends of the tray.	☐
3	Measure and cut the SWA cable, allowing sufficient for termination at each end.	☐
4	Fix the cable to the tray, using cable ties.	☐
5	Cut out a 20mm hole in the metal back box.	☐
6	Remove suitable lengths of outer sleeve and insulation from the cable ends.	☐
7	Dismantle the commando socket and connect the L and N terminals, making sure that they are secure. (Don't forget to place the shroud and socket cable clamp over the cable before connection.)	☐
8	Cut the steel armour strands to length and complete the earth connection. (For correct cpc connection, refer to manufacturers' instructions for the particular type of commando socket you are using.)	☐
9	Reassemble the socket.	☐
10	Fit the banjo and gland to the box and insert the cable. (Don't forget to place the shroud over the cable before insertion into the box.)	☐
11	Make certain that the banjo is securely bolted to the box, and that the cable armour is securely fixed to provide a sound earth connection.	☐

Continued

E19 (CONTINUED)

SWA CABLE INSTALLATION

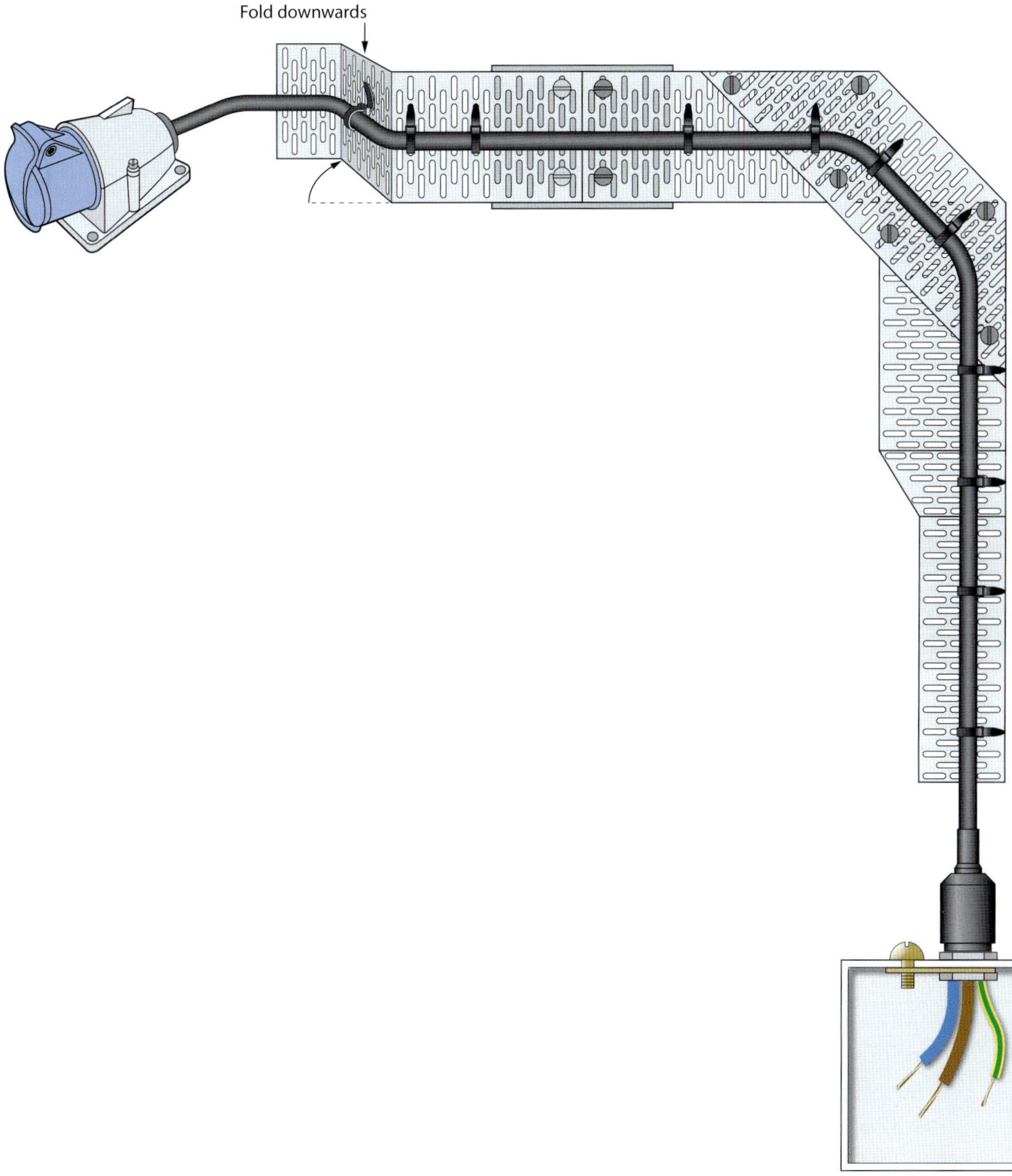

Date completed

Tutor/trainer's feedback/comments

E20
MICC TERMINATION

Your tutor/trainer will provide you with a length of mineral insulated copper clad (MICC) cable, a metal box, a set of termination tools, a gland, a sealing pot, compound and shroud.

Note: there are numerous instruction sets on MICC termination, including City & Guilds SmartScreen, and the City & Guilds textbook *Level 3 NVQ Diploma in Electrotechnical Technology*, Units 305–306. You may find it helpful to refer to these instructions during this task.

Item	Complete all the tasks below	Candidate check ✓
1	Fix the box to a surface.	☐
2	Make certain that the cable end is undamaged, with no signs of damp insulating material. If necessary, obtain an undamaged end by using a hacksaw to cut off about 20mm.	☐
3	Cut and fit the shroud over the cable.	☐
4	Remove a suitable length of outer sheath, allowing for 50mm of bare conductor in the finished product.	☐
5	Use a stripping tool and pliers to remove 50mm of copper sheath, to reveal 50mm of bare conductor.	☐
6	Fit the gland to the cable, but do not tighten.	☐
7	Fix the sealing pot to the cable end using a suitable pot wrench. When fitted, tap the end of the pot to ensure there is no debris left inside.	☐
8	Fit a sealing disc over the conductors, pressing it to the bottom of the pot where required, and insert the sealing compound.	☐
9	Place a second sealing disc over the conductors and secure the assembly using a crimper.	☐
10	Insert the completed end into the metal box, secure the gland, and fit the shroud.	☐
11	Fit sleeving over the bare conductors.	☐
12	Prove that the termination is sound by applying a 500V insulation resistance test between the conductors, and between each conductor and sheath. Note: because you have not terminated the other end of the cable, make certain that the conductors are not touching at that end.	☐
13	Fix the box to a surface.	☐

Continued

E20 (CONTINUED)

MICC TERMINATION

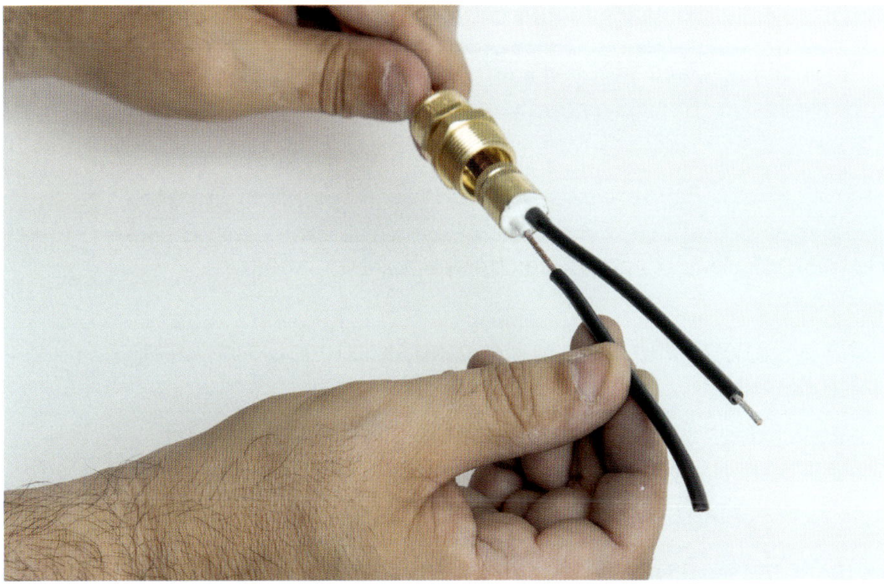

Date completed

Tutor/trainer's feedback/comments

E21

FP200 TERMINATION

Connect a manual fire call point using FP200 cable. The call point is to provide resistive signalling.

Note: the illustration below gives a typical example of how signalling may be configured; however, refer to the instructions provided with the specific call points that you are using for your installation.

Item	Complete all the tasks below	Candidate check ✓
1	Fix the back boxes to a surface, 1m apart. Fit a gland to each box.	☐
2	Cut a length of FP200 cable to 1.4m and fix to the surface, using cable clips.	☐
3	Pass the fixings over the cable ends and pass the cables into the call point back boxes.	☐
4	Secure the gland at each end.	☐
5	Strip the outer sheath at each end of the cable to within 10mm of the cable entry point.	☐
6	Connect the call point at each end such that it would provide resistive signalling to a fire alarm control panel.	☐
7	Prove the circuit operation by connecting a low resistance ohmmeter across the cables at the 'Fire Control Panel' end of the circuit, and operating the call point using a test key.	☐

Continued

E21 (CONTINUED)

FP200 TERMINATION

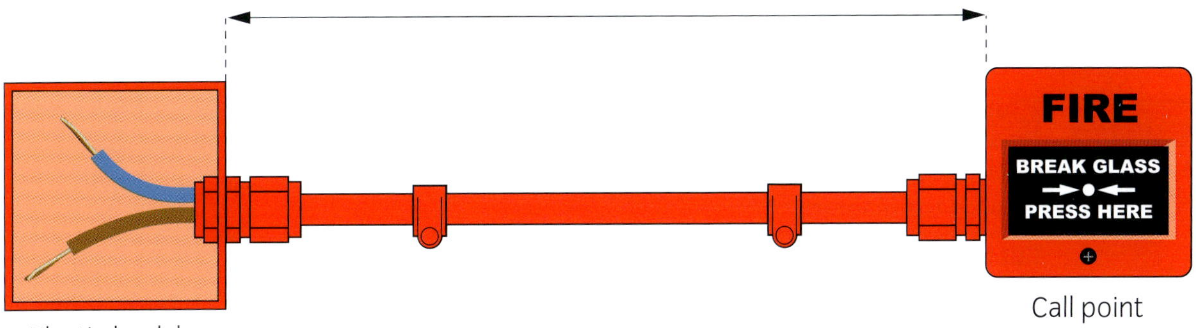

Plastic back box
(Simulating a fire control panel))

Call point
(Mounted on back box)

Connect T1–T4 for 0 Ω switching

Connect T1–T2 for 470 Ω monitoring

Connect T1–T3 for 680 Ω monitoring

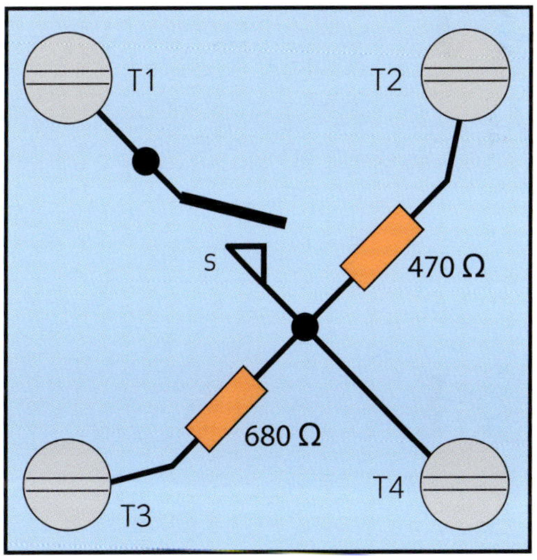

Date completed

Tutor/trainer's feedback/comments

E22
PVC TRUNKING FABRICATION

Accurately measure, cut and drill PVC trunking, and fabricate a 90° bend.

Note 1: when cutting larger trunking, it is helpful to pack the trunking next to the cutting point with wood to prevent the sides from flexing whilst cutting.

Note 2: it is important to cut the trunking and the lid separately as, often, the lid needs to be a different size or shape.

Item	Complete all the tasks below	Candidate check ✓
1	Mark out and cut the trunking lengths using the measurements given in the diagram.	☐
2	Cut the two mitre ends, either by using a mitre box, or by marking the 45° angle using a bevel, and cutting along the line.	☐
3	De-burr the cut ends as required.	☐
4	Measure and mark the positions for the two holes.	☐
5	Drill 20mm holes in each side of the trunking, with the holes directly opposite each other.	☐
6	Fix the trunking to a surface, fit the mitred covers, and prove the 90° angle using a square.	☐
7	Dismantle the assembly once your tutor/trainer has confirmed it is satisfactory.	☐

Continued

E22 (CONTINUED)

PVC TRUNKING FABRICATION

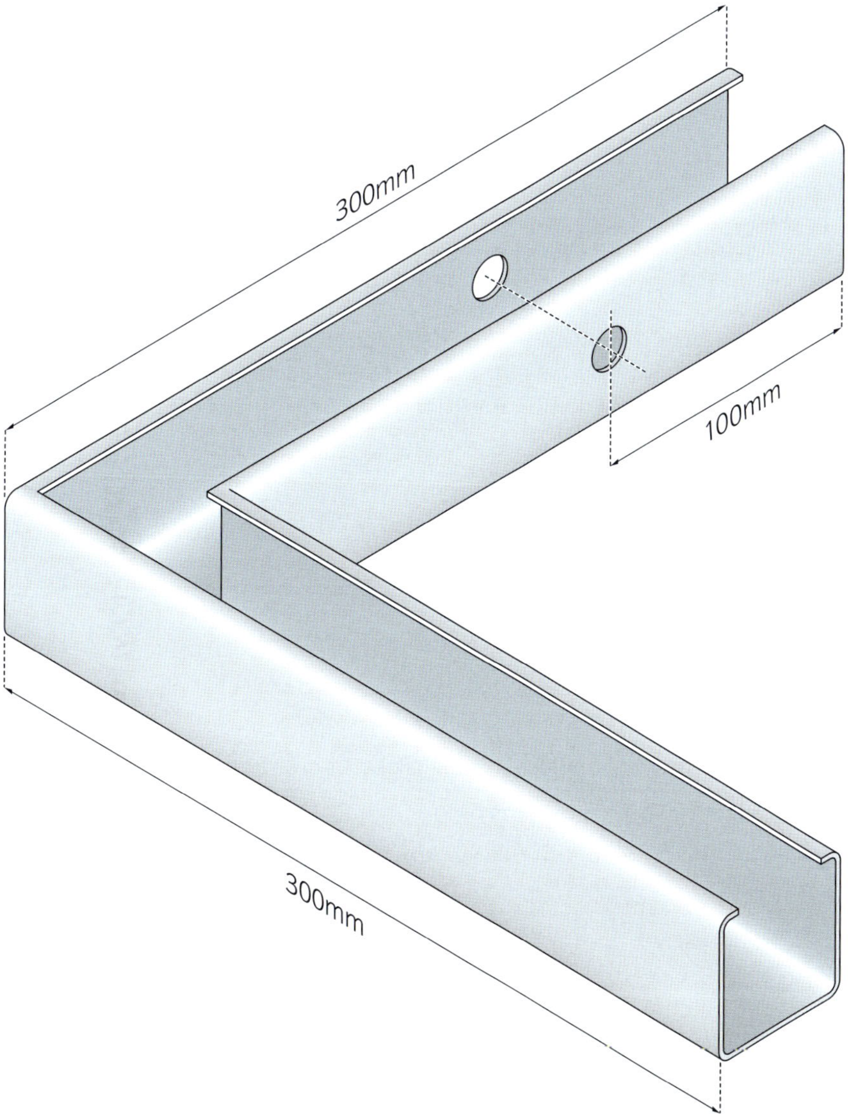

Date completed

Tutor/trainer's feedback/comments

E23

STEEL TRUNKING FABRICATION

Accurately measure, cut and drill steel trunking and fabricate a 90° bend.

Note 1: when cutting or drilling larger trunking, it is helpful to pack the trunking next to the cutting/drilling point with wood to prevent the sides from flexing.

Note 2: it is important to cut the trunking and the lid separately as, often, the lid needs to be a different size or shape.

Note 3: there are numerous instruction sets on forming 90° bends in steel trunking, including on City & Guilds SmartScreen. You may find it helpful to refer to these instructions during this task.

Item	Complete all the tasks below	Candidate check ✓
1	Mark out and cut the trunking lengths using the measurements given in the diagram.	☐
2	Cut and form the mitred joint.	☐
3	De-burr the cut ends as required.	☐
4	Measure and mark the positions for the three holes for the conduit couplers.	☐
5	Drill a 20mm hole for each conduit coupler.	☐
6	Cut two lengths of steel conduit to 50mm length.	☐
7	Ream and de-burr both ends of each piece of conduit.	☐
8	Fit the two pieces of conduit in the positions shown in the diagram using couplers. Also fit an earth tag in the position shown.	☐

Continued

E23 (CONTINUED)

STEEL TRUNKING FABRICATION

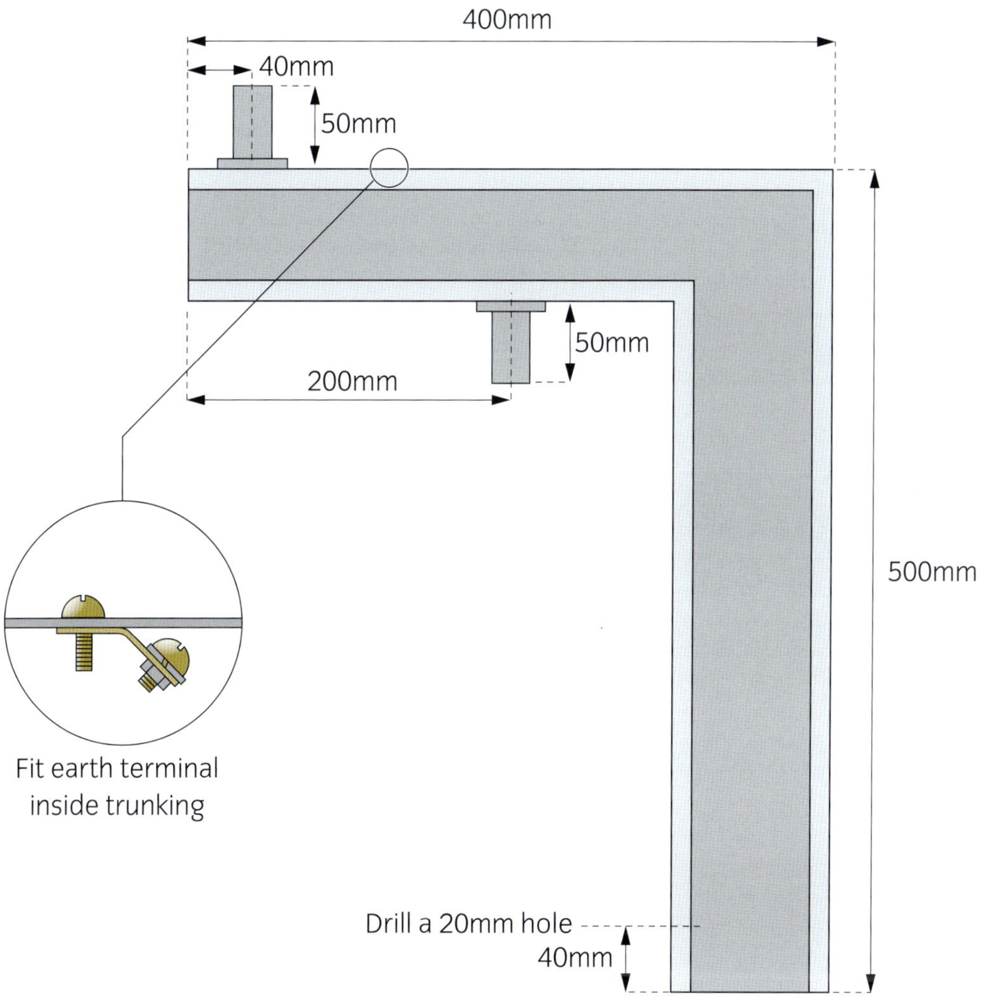

Date completed

Tutor/trainer's feedback/comments

Tasks E25 to E32 are designed to combine to make one single installation. Some of these tasks will make use of assemblies from earlier tasks to reduce material costs, however, at the tutor/trainer's discretion, these assemblies could be constructed again.

It must be remembered that it is not the intention to necessarily complete every task in this handbook, and your tutor/trainer may decide to omit some of these combined tasks due to time constraints, limitation of space, or because the work has been covered in others parts of your course.

The diagram below illustrates the complete installation. Your tutor/trainer may wish to alter the dimensions or layout to suit the available space.

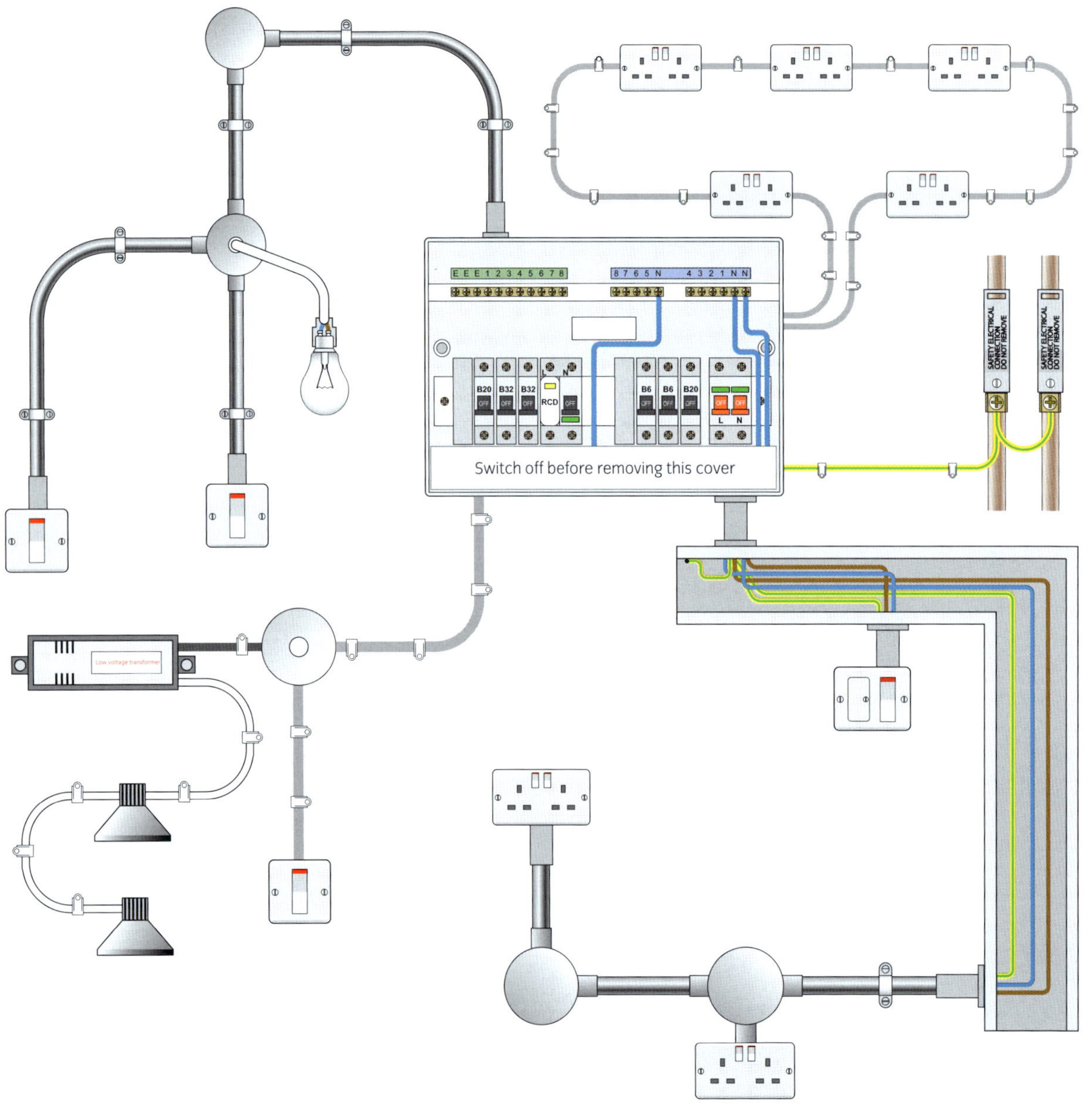

E24

PLANNING AN INSTALLATION

1 MATERIALS LIST

Referring to the Combined tasks installation diagram on the previous page, make a list of the materials that will be required. Include all fixings, couplings, and so on.

Continued

E24 (CONTINUED)

PLANNING AN INSTALLATION

Continued

E24 (CONTINUED)

PLANNING AN INSTALLATION

2 TOOLS LIST

Referring to the Combined tasks installation diagram on page 52, make a list of the tools and PPE that will be required.

Date completed

Tutor/trainer's feedback/comments

E25

SYSTEM INSTALLATION: RING FINAL CIRCUIT

Install a ring final circuit using 2.5mm^2 flat profile twin and cpc cable clipped to a surface.

Before commencing, agree the circuit dimensions with your tutor/trainer and write these on the diagram below.

Make a list of the materials that you will require to complete this task. You may find it helpful to refer to the list that you made in task E24.

Continued

E25 (CONTINUED)

SYSTEM INSTALLATION: RING FINAL CIRCUIT

Item	Complete all the tasks below	Candidate check ✓
1	Agree the cable entry point and method for the consumer unit with your tutor.	☐
2	Determine which circuit breaker you should use. Write the reasons for your choice below. ______________________________	☐
3	Remove knock-outs from metal boxes, where required, and insert rubber grommets.	☐
4	Fix the consumer unit to a surface, after removing any required knock-outs.	☐
5	Fix the socket back boxes to the surface, using the dimensions in the diagram below.	☐
6	Measure and cut the cables to the required lengths. Allow sufficient length at each end for comfortable connection.	☐
7	Clip cables to surface as per the layout diagram below, paying particular attention to the radius of the bends. Ensure that the cables lay flat on the surface at all points. Note: the number and position of the clips will depend on your layout and dimensions. Refer to the IET On-site Guide for guidance on bend radius dimensions and clip spacing.	☐
8	Remove a suitable length of outer sheath on each of the cable ends.	☐
9	Remove the insulation from each conductor, sufficient to fit into the terminals without excess bare copper being exposed.	☐
10	Fit green/yellow sleeve over each cpc conductor.	☐
11	Correctly connect the circuit to form a ring final circuit (do not connect at the consumer unit at this point).	☐
12	Ensure that all terminations are correct, and that they are electrically and mechanically sound.	☐
13	Fit the socket-outlet plates to the back boxes.	☐
14	Referring to the IET publication *Guidance Note 3: Inspection and Testing*, perform the three-step test for proving continuity of a ring final circuit.	☐
15	Record the value of $R_1 + R_2$ below. $R_1 + R_2 =$ ______	☐
16	Complete the connections at the consumer unit.	☐

Continued

E25 (CONTINUED)

SYSTEM INSTALLATION: RING FINAL CIRCUIT

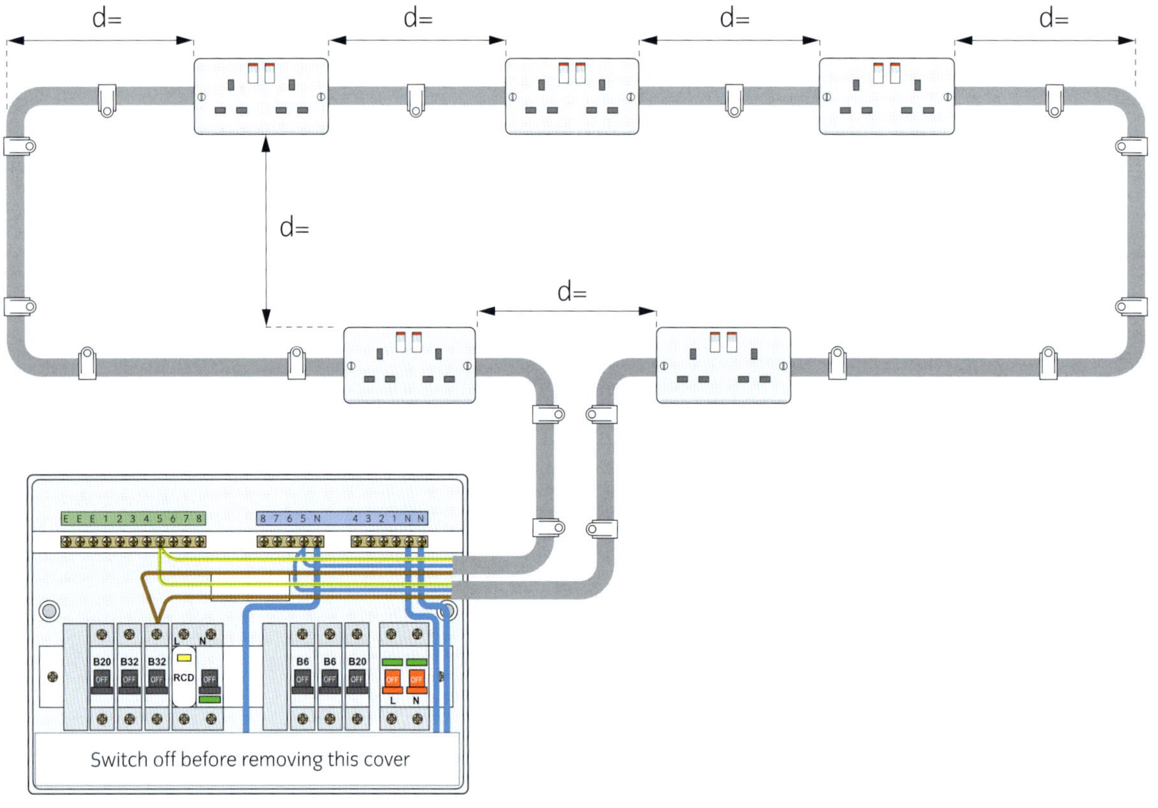

Date completed

Tutor/trainer's feedback/comments

E26

SAFE ISOLATION AND LOCKING-OFF: SINGLE PHASE

Assume that the consumer unit in task E25 is connected to an incoming supply, and that you need to replace one of the sockets in the ring final circuit. You would first need to ensure that the circuit was safe to work on.

The importance of safe isolation and lock-off procedure cannot be emphasised enough. Have you learned it yet? Prove it. Write the procedure here.

Continued

E26 (CONTINUED)

SAFE ISOLATION AND LOCKING-OFF: SINGLE PHASE

Item	Complete all the tasks below	Candidate check ✓
1	Ask your tutor to check your written procedure before proceeding.	☐
2	Isolate the circuit and use an appropriate locking-off kit to secure it.	☐
3	Following the correct method, use an appropriate test instrument to prove that the circuit is dead.	☐
4	Assuming that you have now completed the socket replacement, reinstate the supply.	☐

Date completed

Tutor/trainer's feedback/comments

E27

SYSTEM INSTALLATION: TWO-WAY LIGHTING CIRCUIT

Install a two-way lighting circuit using 1.5mm^2 single core cables in PVC conduit fixed to a surface.

Before commencing, agree the circuit dimensions given in the diagram with your tutor/trainer. Include any changes on the diagram.

Make a list of the materials that you will require to complete this task. You may find it helpful to refer to the list that you made in task E24.

Continued

E27 (CONTINUED)

SYSTEM INSTALLATION: TWO-WAY LIGHTING CIRCUIT

Item	Complete all the tasks below	Candidate check ✓
1	Agree the cable entry point and method for the consumer unit with your tutor.	☐
2	Determine which circuit breaker you should use. Write the reasons for your choice below. ______	☐
3	Fix the elbow, ceiling rose and back boxes to the surface, using the dimensions in the diagram.	☐
4	Measure and cut the four conduit lengths, bending two of them to 90°. De-burr as required. Note: the number and position of supports will depend on your layout and dimensions. Refer to the IET On-site Guide for guidance on bend radius dimensions and support requirements.	☐
5	Measure and cut the line, neutral and cpc cables to the required lengths. Allow sufficient length at each end for comfortable connection.	☐
6	Assemble the conduit using appropriate couplers, fixing to the surface using saddles.	☐
7	Pull the cables through the conduit, using a draw tape if necessary.	☐
8	Remove the insulation from each conductor, sufficient to fit into the terminals without excess bare copper being exposed.	☐
9	Connect the ceiling rose and switches (assume there is no loop out cable). Refer to the connection diagram in task E14 if you are not sure how to do this. Make sure that terminations are electrically and mechanically sound.	☐
10	Fit the elbow cover and ceiling rose cover, and fix the switches to the back boxes.	☐
11	Referring to the IET publication *Guidance Note 3: Inspection and Testing*, perform test method 1 for circuits to prove circuit continuity.	☐
12	Record the value of $R_1 + R_2$ below. $R_1 + R_2 =$ ______	☐
13	Referring to the IET publication *Guidance Note 3: Inspection and Testing*, perform an insulation resistance test between all conductors (make certain that no lamp is fitted during this test).	☐
14	Complete the connections at the consumer unit, making certain that correct sequencing is maintained for the neutral and cpc conductors.	☐

Continued

E27 (CONTINUED)

SYSTEM INSTALLATION: TWO-WAY LIGHTING CIRCUIT

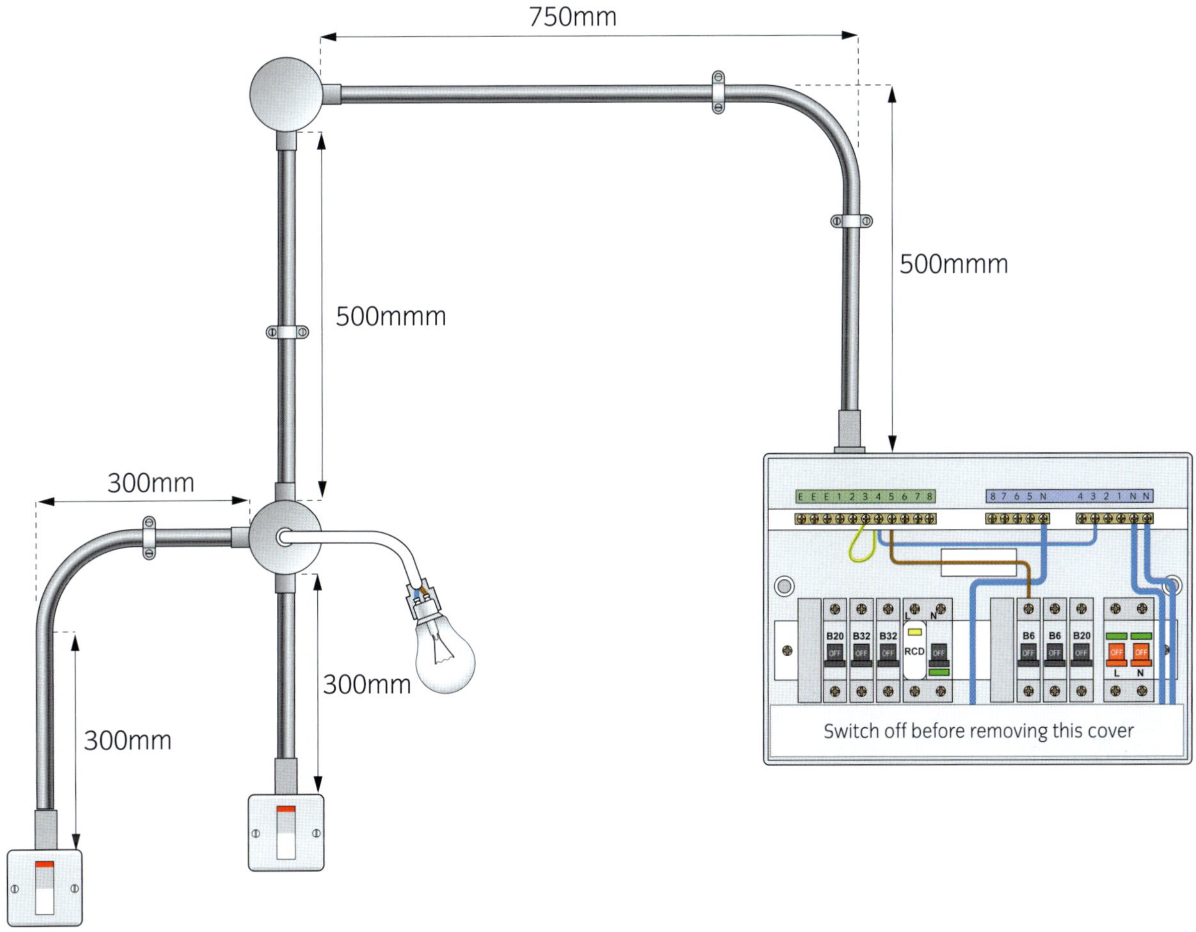

Date completed

Tutor/trainer's feedback/comments

E28

SYSTEM INSTALLATION: STEEL TRUNKING INSTALLATION

Install a circuit containing a 32A isolator, using single-core cables in steel trunking fixed to a surface. Live conductors = $4mm^2$; cpc = $2.5mm^2$.

You may use the trunking assembly that you made in task E23. However, for further practice, your tutor/trainer may ask you to cut and assemble a new trunking section.

Make a list of the materials that you will require to complete this task. You may find it helpful to refer to the list that you made in task E24.

Continued

E28 (CONTINUED)

SYSTEM INSTALLATION: STEEL TRUNKING INSTALLATION

Item	Complete all the tasks below	Candidate check ✓
1	Agree the cable entry point for the consumer unit with your tutor.	☐
2	Determine which circuit breaker you should use. Write the reasons for your choice below.	☐
3	Cut a hole in the consumer unit, as required, to connect the conduit.	☐
4	Fix the trunking assembly to the surface with the conduit coupler aligned with the consumer unit fixing hole.	☐
5	Fit the back box for the fused connection unit as shown in the diagram.	☐
6	Fit the conduit and couplers to the consumer unit and back box.	☐
7	Measure and cut the line, neutral and cpc cables to the required lengths. Allow sufficient length at each end for comfortable connection. Include a cpc conductor to connect the trunking earth terminal.	☐
8	Pass the cables through the trunking and conduits.	☐
9	Remove the insulation from each conductor, sufficient to fit into the terminals without excess bare copper being exposed.	☐
10	Connect the fused connection unit (assume there is no load cable at this stage of the installation). Make sure that terminations are electrically and mechanically sound.	☐
11	Fit the FCU to the back box.	☐
12	Connect the earth conductor to the trunking earth terminal. Connect the other end of this conductor to the appropriate earth terminal in the consumer unit. (This will depend on which circuit you are using in the consumer unit.)	☐
13	Complete the connections at the consumer unit.	☐
14	Fit the trunking covers.	☐

Continued

E28 (CONTINUED)

SYSTEM INSTALLATION: STEEL TRUNKING INSTALLATION

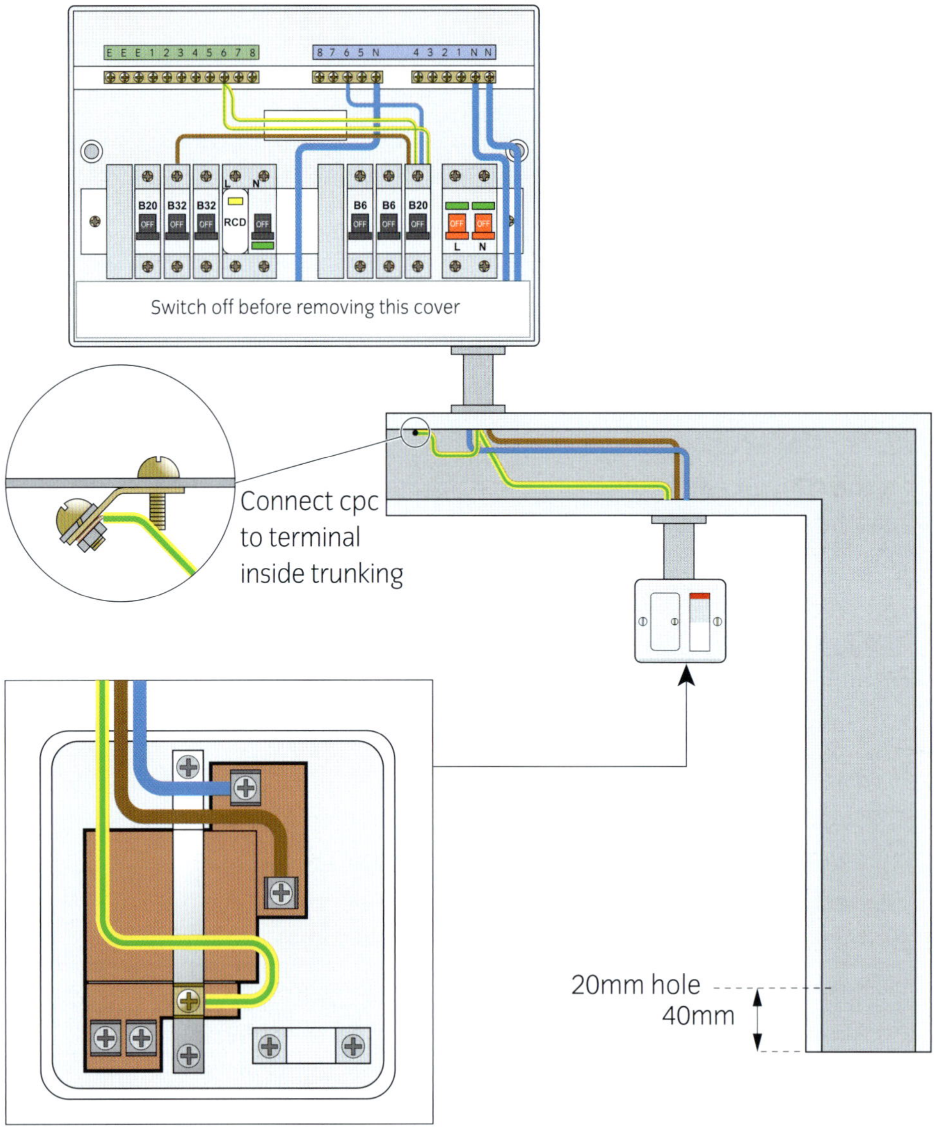

Date completed

Tutor/trainer's feedback/comments

E29

SYSTEM INSTALLATION: RADIAL CIRCUIT IN STEEL CONDUIT

Install a radial circuit containing two 13A socket-outlets, using 2.5mm^2 single core cables in steel trunking fixed to a surface.

You will add this installation to the steel trunking assembly in task E28.

For this task, assume that the consumer unit is live and, therefore, you will have to first perform a safe isolation, locking-off and circuit testing before commencing work.

Item	Complete all the tasks below	Candidate check ✓
1	Determine which circuit breaker you should use. Write the reasons for your choice below. ____________________________________	☐
2	Perform safe isolation, locking-off, and testing of the proposed circuit. Do not continue with the installation until your tutor/trainer has confirmed that you have done this correctly.	☐
3	Fix the socket back boxes and elbows to a surface, using the measurements given in the diagram.	☐
4	Cut the conduit to the required lengths.	☐
5	Ream and de-burr, as required.	☐
6	Thread all conduit ends.	☐
7	Install the conduit as shown in the diagram.	☐
8	Pass the line, neutral and cpc cables through the conduit and trunking and cut to the required lengths. Allow sufficient length for comfortable connection.	☐
9	Fit the elbow covers.	☐
10	Remove the insulation from each conductor, sufficient to fit into the terminals without excess bare copper being exposed.	☐
11	Connect the line, neutral and cpc at the socket outlets, making sure that terminations are electrically and mechanically sound.	☐
12	Fit the socket outlets and trunking covers.	☐
13	Referring to the IET publication *Guidance Note 3: Inspection and Testing*, perform test method 2 for circuits to prove continuity of the cpc conductors. Record the value for R_2 in the space below. R_2 = ______	☐

Continued

E29 (CONTINUED)

SYSTEM INSTALLATION: RADIAL CIRCUIT IN STEEL CONDUIT

Item	Complete all the tasks below	Candidate check ✓
14	Referring to the IET publication *Guidance Note 3: Inspection and Testing*, perform test method 2 for circuits to prove the integrity of the earthing conductors for the steel trunking.	☐
15	Referring to the IET publication *Guidance Note 3: Inspection and Testing*, perform an insulation resistance test between all conductors in the new circuit.	☐
16	Connect the cables in the consumer unit.	☐
17	Remove the lock-off device.	☐

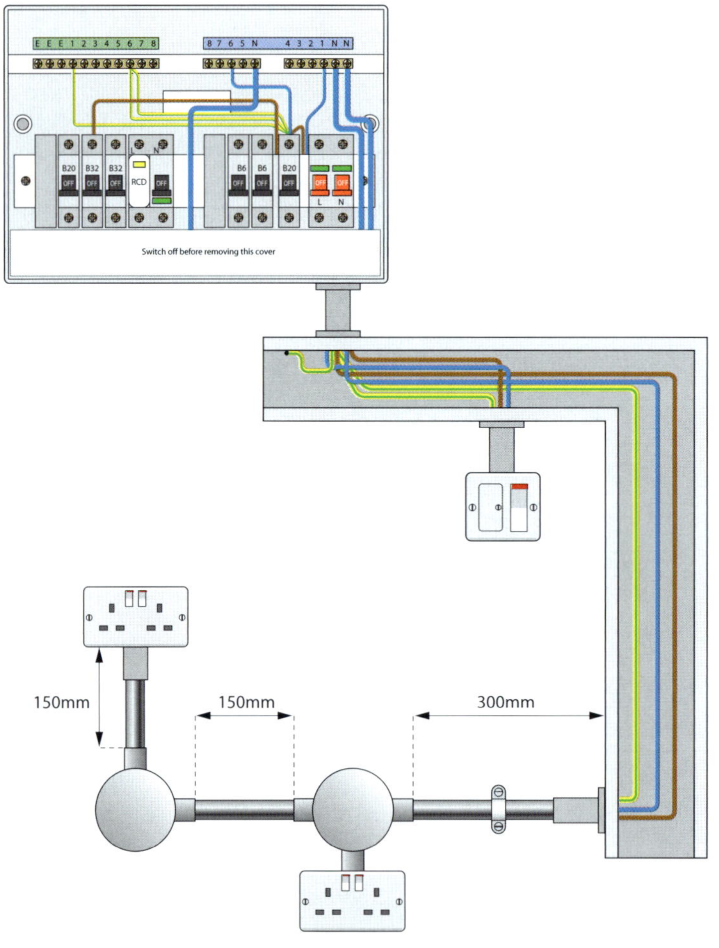

Date completed

Tutor/trainer's feedback/comments

E30

SYSTEM INSTALLATION: PROTECTIVE BONDING

Bond two copper pipes using a protective bonding conductor clipped to a surface.

Agree the CSA of cable to be used with your tutor/trainer before starting this task.

Item	Complete all the tasks below	Candidate check ✓
1	Select suitable bonding clamps for the pipework.	☐
2	Clean and prepare the pipework for installation of clamps.	☐
3	Fit and secure the bonding clamps.	☐
4	Clip the cable to the surface.	☐
5	Strip the insulation to allow a continuous circuit between the clamps.	☐
6	Secure the cable in the bonding clamp lugs, and ensure that all labels are secure.	☐
7	Connect the cable in the consumer unit.	☐
8	Ensure that all terminations are electrically and mechanically sound.	☐
9	Referring to the IET publication *Guidance Note 3: Inspection and Testing*, perform test method 2 for circuits to prove the integrity of the earth bonding conductors.	☐

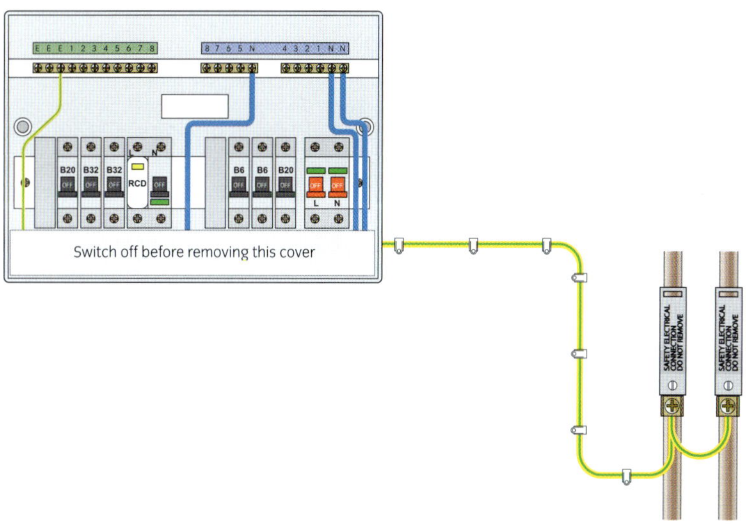

Date completed

Tutor/trainer's feedback/comments

E31

SYSTEM INSTALLATION: TRANSFORMER AND EXTRA-LOW-VOLTAGE LIGHTING

Install a 12V dc extra-low-voltage lighting system, connected on its own circuit.

Agree the cable entry point in the consumer unit and the circuit dimensions with your tutor/trainer before commencing.

For this task, assume that the consumer unit is live and, therefore, you will have to first perform safe isolation, locking-off and suitable tests before commencing work.

Note: the diagram below provides a general design. The actual design will depend on the available transformer and lights. You tutor will provide you with suitable components.

Item	Complete all the tasks below	Candidate check ✓
1	Determine which circuit breaker you should use. Write the reasons for your choice below. ____ ____	☐
2	Perform safe isolation, locking-off and testing of the proposed circuit. Do not continue with the installation until your tutor/trainer has confirmed that you have done this correctly.	☐
3	Fix the junction box, switch back box and transformer to a surface, using the dimensions agreed with your tutor/trainer.	☐
4	On the consumer unit, remove knock-out or cut an entry hole as necessary for the cable entry. Fit a suitable IP-rated grommet.	☐
5	Cut suitable lengths of 1.5mm^2 flat profile twin and cpc cable, allowing sufficient at each end for connection.	☐
6	Clip the cables to the surface, making sure that the bends are not too tight. Refer to the IET On-site Guide for guidance on bend radius dimensions and clip spacing.	☐
7	Fix the extra-low-voltage cable and lamp units to the surface, using suitable methods. (Note: this will depend on the components used.)	☐
8	Remove a suitable length of outer sheath on each of the cable ends.	☐
9	Remove the insulation from each conductor, sufficient to fit into the terminals without excess bare copper being exposed.	☐
10	Fit green/yellow sleeve over all cpc conductors on all 230V parts of the circuit. Also ensure that live conductors in the switch circuit are correctly identified.	☐
11	Connect the cables in the junction box and switch. **Do not connect the transformer!**	☐

Continued

E31 (CONTINUED)

SYSTEM INSTALLATION: TRANSFORMER AND EXTRA-LOW-VOLTAGE LIGHTING

You will now perform an insulation resistance test on the circuit. Explain, in the space below, why the transformer would not be connected at this stage, and the possible consequences of performing this test with the transformer connected.

Item	Complete all the tasks below	Candidate check ✓
12	Perform insulation resistance testing on all conductors in the circuit. Record your results in the table below.	☐
13	Complete the transformer and consumer unit connections.	☐
14	Fit the junction box cover and the light switch plate.	☐
15	Remove locking-off device and simulate energising of the circuit.	☐

Applied test	Result
Neutral – earth	
Line – neutral	
Line – earth	

Continued

E31 (CONTINUED)

SYSTEM INSTALLATION: TRANSFORMER AND EXTRA-LOW-VOLTAGE LIGHTING

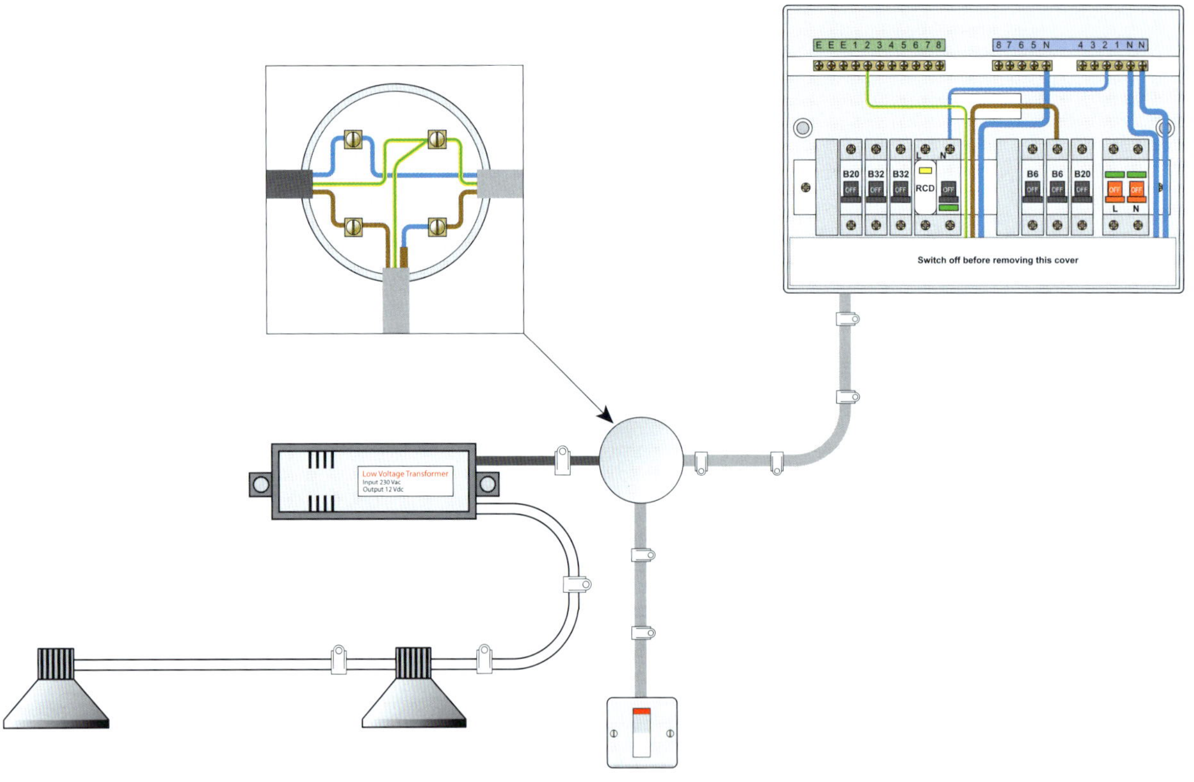

Date completed

Tutor/trainer's feedback/comments

E32

SYSTEM INSTALLATION: INCOMING SUPPLY (SIMULATED)

Connect incoming supply tails in a single-phase consumer unit.

Important: Your tutor/trainer will have some means in place of ensuring that the supply you are connecting to is isolated and locked off. However, never assume that a circuit is dead simply because someone tells you that it is – not even your tutor/trainer! Before beginning this task, confirm with your tutor/trainer that:

- the circuit is dead
- the circuit cannot be energised whilst you are working on it.

Determine the correct cross-sectional area (CSA) cable for the line, neutral and earthing conductors for the incoming supply. Write the cable sizes below, giving reasons for your choice.

Line:

Neutral:

Earthing conductor:

Item	Complete all the tasks below	Candidate check ✓
1	Measure and cut the three cables to the required lengths.	☐
2	Pass the cables from the consumer unit to the termination point, fixing to a surface as required, using suitable means. Refer to IET On-site Guide for guidance on cable fixing.	☐
3	Remove all cable sheathes and insulation to reveal suitable lengths of bare conductor.	☐
4	Connect all cables at both ends, making certain that terminations are electrically and mechanically sound.	☐

Continued

E32 (CONTINUED)

SYSTEM INSTALLATION: INCOMING SUPPLY (SIMULATED)

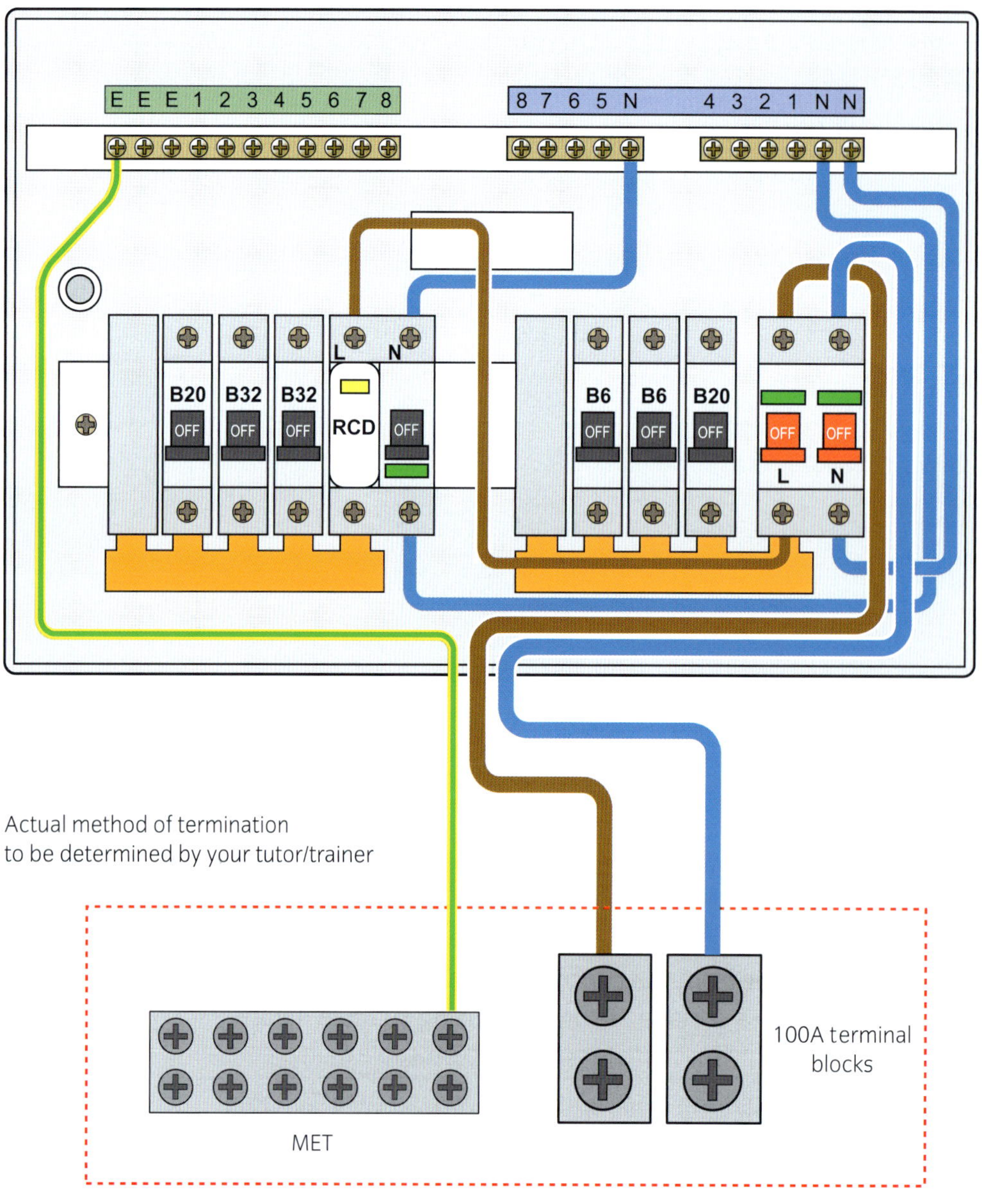

Date completed

Tutor/trainer's feedback/comments

E33

INITIAL VERIFICATION TESTING: SINGLE-PHASE SYSTEM

Perform initial verification inspection and testing on a single-phase installation, and complete all initial verification forms.

The combined tasks E25–E32 provided you with a simulated installation on which you can practise testing.

If you have completed all of the tasks E25–E32, then your installation should look something like that in the illustration below. However, it might well be the case that for any number of reasons you have had to deviate from this specification. But this should not prevent you from completing this task; simply complete the initial verification forms for your installation.

It is not expected that you will complete this task in just one single lesson period. The intention of this task is to provide you with a platform on which to learn testing, and this may even be the first opportunity you have to actually use the test instruments. You may wish to complete this task working with another learner.

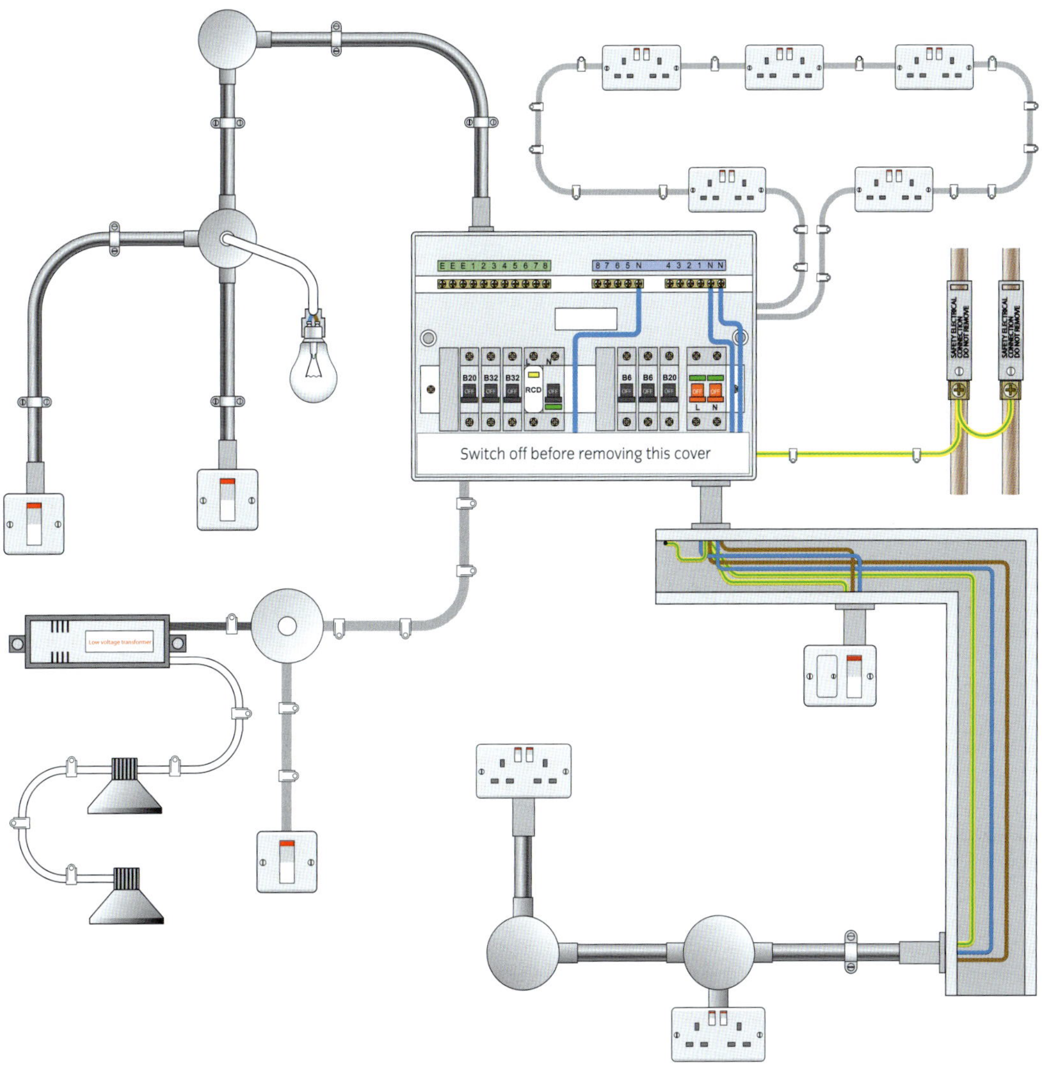

Continued

E33 (CONTINUED)

INITIAL VERIFICATION TESTING: SINGLE-PHASE SYSTEM

Testing must be carried out in accordance with BS 7671, where the prescribed tests are detailed in the sequence in which they must be performed.

Details of how the tests are carried out are given in the IET publication *Guidance Note 3 – Inspection and Testing*. However, detailed instructions are also provided in the City & Guilds textbook *Level 3 NVQ Diploma in Electrotechnical Technology*, Units 307–308, Outcome 4.

Referring either to the BS 7671 Regulations, the City & Guilds textbook, or other sources, in the space below write down the tests that you will need to carry out for your installation, in the order that you will perform them.

Ask your tutor/trainer to check your list before proceeding.

Continued

E33 (CONTINUED)

INITIAL VERIFICATION TESTING: SINGLE-PHASE SYSTEM

Complete the initial verification forms provided in this task. The following notes relating to each part may help you. However, these are not intended to provide detailed instructions on how to perform the actual tests; such instructions may be found in the IET *Guidance Note 3 – Inspection and Testing*, the City & Guilds textbook *Level 3 NVQ Diploma in Electrotechnical Technology*, SmartScreen, or your course notes.

TOP SECTION:

DB Ref No	This should be displayed on the cover of the distribution board.
Location	Enter the physical location of the distribution board.
Z_s at DB (Z_{db})	For a DB connected directly to the main incoming supply, this would be the Ze measurement. However, when measured at a sub-main board, the measurement is taken in the same manner only it is referred to as Z_s at DB.
I_{pf} at DB	This test has to be performed live, using a suitable test instrument. However, the information may be obtained 'by enquiry' with the electricity supply company. Your tutor/trainer must be present whilst you are performing this test on a live circuit.
Correct supply polarity	This should be verified by live testing at the incoming supply using a suitable voltage test instrument.
Phase sequence	This test applies only to three-phase supplies and so can be ignored for this task.
Circuits vulnerable to damage	List any circuits in the installation that contain equipment/devices that may be damaged during testing. (These must be disconnected and bridged prior to testing.)
Details of test instruments	Make and serial number of test instrument(s) used for this testing.

Continued

E33 (CONTINUED)

INITIAL VERIFICATION TESTING: SINGLE-PHASE SYSTEM

COLUMNS 1–22:

Item	Complete all the tasks below	Candidate check ✓
1	The circuit number in the DB.	☐
2	Add a brief description (eg 'downstairs lighting').	☐
3	The BS EN number that defines the type of protective device for each circuit (eg BS EN60898). This information is found on the circuit breaker.	☐
4	The type of circuit breaker refers to the fault current at which the device will operate, breaking the circuit within 0.1 seconds. This is denoted by the letters B, C, or D. This information is found on the circuit breaker.	☐
5	The rating of the circuit breaker refers to the normal (no fault condition) maximum current that the device will pass in normal service.	☐
6	The breaking capacity is the maximum current that the device will withstand before catastrophic failure. Measured in kA. This information is found on the circuit breaker.	☐
7	This refers to the method of installation/containment of the cables, which has a bearing on the maximum current carrying capacity. This information is found in BS 7671, Table 4A2.	☐
8	The cross-sectional area of the live conductors.	☐
9	The cross-sectional area of the cpc.	☐
10	This applies to ring-final circuits only. r_1 is the end-to-end resistance of the line conductor.	☐
11	This applies to ring-final circuits only. r_n is the end-to-end resistance of the neutral conductor.	☐
12	This applies to ring-final circuits only. r_2 is the end-to-end resistance of the cpc.	☐
13	This applies to all circuits. However, the test methods vary between lighting/radial circuits and ring-final circuits. For details on performing these tests, refer to *IET Guidance Note 3 – Inspection and Testing*.	☐
14	This applies to all circuits. This test is not required if $R_1 + R_2$ has been measured and recorded in column 13.	☐

Continued

E33 (CONTINUED)

INITIAL VERIFICATION TESTING: SINGLE-PHASE SYSTEM

Item	Complete all the tasks below	Candidate check ✓
15	**Important:** Before performing insulation resistance testing, the supply must be isolated, all items of equipment should be disconnected from sockets, light fittings, etc. All cpc and bonding conductors must be connected. All breakers must be set to ON. Test for two-way lighting circuits must be performed with the switches in both positions.	☐
16	The insulation resistance between the live and cpc conductors (line – cpc, and neutral – cpc).	☐
17	Once correct polarity has been confirmed, place a tick in this column.	☐
18	Having measured Z_{db} (that is; Z_s at DB) and $R_1 + R_2$, the total earth fault loop impedance $Z_s = Z_{db} + (R_1 + R_2)$. It is also possible to measure this value using an earth fault loop impedance tester, where safe and suitable to do so. Note: the values obtained will vary depending on the external earthing method. Full details of these tests are given in *IET Guidance Note 3 – Inspection and Testing*.	☐
19	The breaking time of the RCD when tested at 1 x rated current.	☐
20	The breaking time of the RCD when tested at 5 x rated current. This test is not required for devices rated >30mA.	☐
21	This is a tick box confirming that the device operates correctly when the test button is pressed.	☐
22	Add comments on circuits, fault conditions (not relevant for an initial inspection), or any other relevant information.	☐

Continued

Date completed

Tutor/trainer's feedback/comments

E33 (CONTINUED)

INITIAL VERIFICATION TESTING: SINGLE-PHASE SYSTEM

Tested by:

The certificate is only valid if this section is completed by the test engineer.

Form 2 Form No:............../2

ELECTRICAL INSTALLATION CERTIFICATE
(REQUIREMENTS FOR ELECTRICAL INSTALLATIONS - BS 7671 [IET WIRING REGULATIONS])

DETAILS OF THE CLIENT

.... Post Code: ....

....

INSTALLATION ADDRESS ....

....

.... Post Code: ....

DESCRIPTION AND EXTENT OF THE INSTALLATION Tick boxes as appropriate	
Description of installation:	New installation ☐
Extent of installation covered by this Certificate:	Addition to an existing installation ☐
(Use continuation sheet if necessary) see continuation sheet No:	Alteration to an existing installation ☐

FOR DESIGN
I/We being the person(s) responsible for the design of the electrical installation (as indicated by my/our signatures below), particulars of which are described above, having exercised reasonable skill and care when carrying out the design hereby CERTIFY that the design work for which I/we have been responsible is to the best of my/our knowledge and belief in accordance with BS 7671:2008, amended to .... (date) except for the departures, if any, detailed as follows:

Details of departures from BS 7671 (Regulations 120.3 and 133.5):

The extent of liability of the signatory or the signatories is limited to the work described above as the subject of this Certificate.

For the DESIGN of the installation: **(Where there is mutual responsibility for the design)

Signature: ....Date: .... Name (IN BLOCK LETTERS): .... Designer No 1

Signature: ....Date: .... Name (IN BLOCK LETTERS): .... Designer No 2**

FOR CONSTRUCTION
I/We being the person(s) responsible for the construction of the electrical installation (as indicated by my/our signatures below), particulars of which are described above, having exercised reasonable skill and care when carrying out the construction hereby CERTIFY that the construction work for which I/we have been responsible is to the best of my/our knowledge and belief in accordance with BS 7671:2008, amended to ....(date) except for the departures, if any, detailed as follows:

Details of departures from BS 7671 (Regulations 120.3 and 133.5):

The extent of liability of the signatory is limited to the work described above as the subject of this Certificate.

For CONSTRUCTION of the installation:

Signature: ....Date: .... Name (IN BLOCK LETTERS): ....

FOR INSPECTION & TESTING
I/We being the person(s) responsible for the inspection & testing of the electrical installation (as indicated by my/our signatures below), particulars of which are described above, having exercised reasonable skill and care when carrying out the inspection & testing hereby CERTIFY that the work for which I/we have been responsible is to the best of my/our knowledge and belief in accordance with BS 7671:2008, amended to ....(date) except for the departures, if any, detailed as follows:

Details of departures from BS 7671 (Regulations 120.3 and 133.5):

The extent of liability of the signatory is limited to the work described above as the subject of this Certificate.

For INSPECTION AND TESTING of the installation:

Signature: ....Date: .... Name (IN BLOCK LETTERS): ....

NEXT INSPECTION
I/We the designer(s), recommend that this installation is further inspected and tested after an interval of not more than .... years/months.

Page....of .....

Continued

E33 (CONTINUED)

INITIAL VERIFICATION TESTING: SINGLE-PHASE SYSTEM

Form 2 Form No: ..../2

PARTICULARS OF SIGNATORIES TO THE ELECTRICAL INSTALLATION CERTIFICATE

Designer (No 1)
Name: .... Company: ....
Address: ....
.... Postcode: ....Tel No: ....

Designer (No 2)
(if applicable)
Name: .... Company: ....
Address: ....
.... Postcode: ....Tel No: ....

Constructor
Name: .... Company: ....
Address: ....
.... Postcode: ....Tel No: ....

Inspector
Name: .... Company: ....
Address: ....
.... Postcode: ....Tel No: ....

SUPPLY CHARACTERISTICS AND EARTHING ARRANGEMENTS Tick boxes and enter details, as appropriate

Earthing arrangements	Number and Type of Live Conductors	Nature of Supply Parameters	Supply Protective Device Characteristics
TN-C ☐ TN-S ☐ TN-C-S ☐ TT ☐ IT ☐	a.c. ☐ d.c. ☐ 1-phase, 2-wire ☐ 2-wire ☐ 1-phase, 3-wire ☐ 3-wire ☐ 2-phase, 3-wire ☐ other ☐ 3-phase, 3-wire ☐ 3-phase, 4-wire ☐	Nominal voltage, $U/U_0^{(1)}$ .... V Nominal frequency, $f^{(1)}$ .... Hz Prospective fault current, $I_{pf}^{(2)}$ .... kA External loop impedance, $Z_e^{(2)}$ ....Ω	Type: .... Rated current....... A
Other sources ☐ of supply (to be detailed on attached schedules)	Confirmation of supply polarity ☐	*(Note: (1) by enquiry, (2) by enquiry or by measurement)*	

PARTICULARS OF INSTALLATION REFERRED TO IN THE CERTIFICATE Tick boxes and enter details, as appropriate

Means of Earthing

Distributor's facility ☐

Installation earth electrode ☐

Maximum Demand

Maximum demand (load) .... kVA / Amps Delete as appropriate

Details of Installation Earth Electrode *(where applicable)*

Type (e.g. rod(s), tape, etc.)	Location	Electrode resistance to Earth
....	....	....................................Ω

Main Protective Conductors

Earthing conductor: material .... csa ....mm^2 Continuity and connection verified ☐

Main protective bonding conductors material .... csa ....mm^2 Continuity and connection verified ☐

To incoming water and/or gas service ☐ To other elements: ....

Main Switch or Circuit-breaker

BS, Type and No. of poles .... Current rating ....A Voltage rating V

Location .... Fuse rating or setting...................... A

Rated residual operating current $I_{\Delta n}$ = ... mA, and operating time of .... ms (at $I_{\Delta n}$) (applicable only where an RCD is suitable and is used as a main circuit breaker)

COMMENTS ON EXISTING INSTALLATION (in the case of an addition or alteration see Section 633):
....
....
....
....

SCHEDULES
The attached Schedules are part of this document and this Certificate is valid only when they are attached to it.
.... Schedules of Inspections and .... Schedules of Test Results are attached.
(Enter quantities of schedules attached)

Page ....of....

Continued

E33 (CONTINUED)

INITIAL VERIFICATION TESTING: SINGLE-PHASE SYSTEM

Form 3   Form No: ..../3

SCHEDULE OF INSPECTIONS (for new installation work only)

Methods of protection against electric shock

Both basic and fault protection:

- ☐ (i) SELV (note 1)
- ☐ (ii) PELV
- ☐ (iii) Double insulation
- ☐ (iv) Reinforced insulation

Basic protection: (note 2)

- ☐ (i) Insulation of live parts
- ☐ (ii) Barriers or enclosures
- ☐ (iii) Obstacles (note 3)
- ☐ (iv) Placing out of reach (note 4)

Fault protection:

(i) Automatic disconnection of supply:

- ☐ Presence of earthing conductor
- ☐ Presence of circuit protective conductors
- ☐ Presence of protective bonding conductors
- ☐ Presence of supplementary bonding conductors
- ☐ Presence of earthing arrangements for combined protective and functional purposes
- ☐ Presence of adequate arrangements for other sources, where applicable
- ☐ FELV
- ☐ Choice and setting of protective and monitoring devices (for fault and/or overcurrent protection)

(ii) Non-conducting location: (note 5)

- ☐ Absence of protective conductors

(iii) Earth-free local equipotential bonding: (note 6)

- ☐ Presence of earth-free local equipotential bonding

(iv) Electrical separation: (note 7)

- ☐ Provided for **one item** of current-using equipment
- ☐ Provided for **more than one item** of current-using equipment

Additional protection:

- ☐ Presence of residual current devices(s)
- ☐ Presence of supplementary bonding conductors

Prevention of mutual detrimental influence

- ☐ (a) Proximity to non-electrical services and other influences
- ☐ (b) Segregation of Band I and Band II circuits or use of Band II insulation
- ☐ (c) Segregation of safety circuits

Identification

- ☐ (a) Presence of diagrams, instructions, circuit charts and similar information
- ☐ (b) Presence of danger notices and other warning notices
- ☐ (c) Labelling of protective devices, switches and terminals
- ☐ (d) Identification of conductors

Cables and conductors

- ☐ Selection of conductors for current-carrying capacity and voltage drop
- ☐ Erection methods
- ☐ Routing of cables in prescribed zones
- ☐ Cables incorporating earthed armour or sheath, or run within an earthed wiring system, or otherwise adequately protected against nails, screws and the like
- ☐ Additional protection provided by 30 mA RCD for cables concealed in walls (where required in premises not under the supervision of a skilled or instructed person)
- ☐ Connection of conductors
- ☐ Presence of fire barriers, suitable seals and protection against thermal effects

General

- ☐ Presence and correct location of appropriate devices for isolation and switching
- ☐ Adequacy of access to switchgear and other equipment
- ☐ Particular protective measures for special installations and locations
- ☐ Connection of single-pole devices for protection or switching in line conductors only
- ☐ Correct connection of accessories and equipment
- ☐ Presence of undervoltage protective devices
- ☐ Selection of equipment and protective measures appropriate to external influences
- ☐ Selection of appropriate functional switching devices

Inspected by .... Date ....

NOTES:

✓ to indicate an inspection has been carried out and the result is satisfactory

N/A to indicate that the inspection is not applicable to a particular item

An entry must be made in every box.

1. SELV An extra-low voltage system which is electrically separated from Earth and from other systems. The particular requirements of the Regulations must be checked (see Section 414)
2. Method of basic protection - will include measurement of distances where appropriate
3. Obstacles - only adopted in special circumstances (see Regulations 416.2 and 417.2)
4. Placing out of reach - only adopted in special circumstances (see Regulation 417.3)
5. Non-conducting locations - not applicable in domestic premises and requiring special precautions (see Regulation 418.1)
6. Earth-free local equipotential bonding - not applicable in domestic premises, only used in special circumstances (see Regulation 418.2)
7. Electrical separation (see Section 413 and Regulation 418.3)

Page ..... of ......

Continued

E33 (CONTINUED)

INITIAL VERIFICATION TESTING: SINGLE-PHASE SYSTEM

Form 4

Form No: ..../4

GENERIC SCHEDULE OF TEST RESULTS

DB reference no ....
Location ....
Zs at DB (Ω) ....
I_{pf} at DB (kA) ....
Correct supply polarity confirmed ☐
Phase sequence confirmed (where appropriate) ☐

Details of circuits and/or installed equipment vulnerable to damage when testing ....
....
....
....
....

Details of test instruments used (state serial and/or asset numbers)
Continuity ....
Insulation resistance ....
Earth fault loop impedance ....
RCD ....
Earth electrode resistance ....

Tested by:
Name (Capitals) ....
Signature .... **Date** ....

Circuit details									**Test results**												
		Overcurrent device				Conductor details			Ring final circuit continuity (Ω)			Continuity (Ω) ($R_1 + R_2$) or R_2		Insulation Resistance (MΩ)		Polarity	Z_s (Ω)	RCD (ms)			Remarks (continue on a separate sheet if necessary)
Circuit number 1	Circuit Description 2	BS (EN) 3	type 4	rating (A) 5	breaking capacity (kA) 6	Reference Method 7	Live (mm^2) 8	cpc (mm^2) 9	r_1 (line) 10	r_n (neutral) 11	r_2 (cpc) 12	($R_1 + R_2$) * 13	R_2 14	Live-Live 15	Live-E 16	17	18	@ $I_{\Delta n}$ 19	@ $5I_{\Delta n}$ 20	Test button operation 21	22

* Where there are no spurs connected to a ring final circuit this value is also the ($R_1 + R_2$) of the circuit.

Continued

E33 (CONTINUED)

INITIAL VERIFICATION TESTING: SINGLE-PHASE SYSTEM

Date completed

Tutor/trainer's feedback/comments

E34
TEST RESULTS ANALYSIS

Check the test results obtained in task E33 against relevant regulations and standards. (In practice, you would normally do this as you perform each test or check).

COLUMN 5:

For each circuit, you will check that the current rating of each breaker is less than the maximum current-carrying capacity for the associated cable. For example, in task E33, for the radial circuit, you would find the maximum value of current for 2.5mm² in BS 7671, Table 4D1A. (We will ignore rating factors at this time.)

In the table below, record the current rating for each cable as given in BS 7671.

Circuit	Current rating
1	
2	
3	
4	
5	

COLUMN 6:

Check that the breaking capacity of each circuit breaker is greater than the Prospective Fault Current (I_{pf}). Enter the values into the table below.

i_{pf} value	Breaker 1	Breaker 2	Breaker 3	Breaker 4	Breaker 5

Continued

E34 (CONTINUED)
TEST RESULTS ANALYSIS

COLUMN 13:

Note: because of the short cable lengths used in task E33, it may be difficult to measure $R_1 + R_2$. However, for a practical circuit containing longer cable runs, you would estimate the cable length and then, using the cable resistance value obtained from Table B1 in *IET Guidance Note 3*, calculate the expected resistance value. The measured value of $R_1 + R_2$ should be close to the calculated value.

It would still be a helpful training exercise for you to perform this step for the installation in task E33, even though the cable resistance values will be very low. Enter your calculated and measured values for R_1 and R_2 into the table below.

	Circuit 1	Circuit 2	Circuit 3	Circuit 4	Circuit 5
Calculated value					
Measured value					

COLUMNS 15 AND 16:

The applied voltages and minimum values of insulation resistance are found in BS 7671, Table 61. Enter the minimum values of insulation resistance for each circuit into the table below.

Circuit 1	Circuit 2	Circuit 3	Circuit 4	Circuit 5

COLUMN 18:

For each circuit, check that the measured Z_S is not greater than 80% of the value given in BS 7671, Tables 41.2 to 41.4, or 100% as given in Tables A1 to A4 in *IET Guidance Note 3*. Complete the table below and use this to confirm acceptable values of Z_S.

	Circuit 1	Circuit 2	Circuit 3	Circuit 4	Circuit 5
Tabulated Z_S value					
Tabulated Z_S x 0.8					
Measured Z_S value					

Date completed

Tutor/trainer's feedback/comments

E35

THREE-PHASE DISTRIBUTION BOARD

Replace the single-phase consumer unit in combined tasks E25–E32 with a three-phase distribution board.

This task focuses on the connection of a three-phase distribution board. Your centre may have a pre-installed rig for you to work on. However, this task provides you with an option to replace the single-phase consumer unit in combined tasks E25–E32 with a three-phase distribution board.

Because the three-phase board will be much larger than the single-phase board used for earlier tasks, it is unlikely that you will be able to simply swop out the boards – you will have to move or modify some of your circuits.

If you have completed combined tasks E25–E32 as per this manual, then your modified installation will look something like that illustrated in the diagram included in this task.

Item	Complete all the tasks below	Candidate check ✓
1	Disconnect all conductors from the single-phase consumer unit and remove the board. Make certain that all circuit breakers, covers, screws, etc., are in place before returning the board to your tutor/trainer.	☐
2	If your new board allows suitable entry points, align it such that the ring-final circuit, the ELV lighting circuit, bonding conductor and the steel trunking are left untouched. You will have to withdraw the cables from conduits to enable replacement of the boards. Mark the four entry points on the new board and make suitable cut-outs for cable entry and conduit connection.	☐
3	Adjust the position of the two-way lighting circuit to enable conduit connection to the new board. You may be able to adjust the length of the existing conduit to fit the new board, but you may have to move some of the other components to make space for the new board.	☐
4	Fix the distribution board to the surface. Secure conduit connections. Pass all other conductors into the board.	☐
5	Connect each of the five circuits to suitable circuit breakers, using the following sequence: Ring-final circuit – L1 Two-way lighting circuit – L2 Extra-low-voltage lighting circuit – L2 Isolator switch – L3 Radial socket circuit – L3	☐
6	Ensure that all cpc cables, including the protective bonding conductor, are connected.	☐

Continued

E35 (CONTINUED)

THREE-PHASE DISTRIBUTION BOARD

CONNECTING THE THREE-PHASE INCOMING SUPPLY CABLES:

You will have to remove the original single-phase supply cables and replace these with a suitable three-phase supply. The actual method of providing this supply will be determined by the facilities at your centre, and your tutor/trainer will instruct you on how this is to be carried out.

Important: Your tutor/trainer will have some means in place of ensuring that the supply you are connecting to is isolated and locked off. However, never assume that a circuit is dead simply because someone tells you that it is – not even your tutor/trainer! Before beginning this task, confirm with your tutor/trainer that:

- the circuit is dead
- the circuit cannot be energised whilst you are working on it.

Determine the correct cross-sectional area (CSA) cable for the line, neutral and earthing conductors for the incoming supply. Write the cable sizes below, giving reasons for your choice:

Line:

Neutral:

Earthing conductor:

Item	Complete all the tasks below	Candidate check ✓
7	Measure and cut the cables to the required lengths.	☐
8	Pass the cables from the distribution board to the termination point, fixing to a surface as required using suitable means. Refer to the IET On-site Guide for guidance on cable fixing.	☐
9	Remove all cable sheaths and insulation to reveal suitable lengths of bare conductor.	☐
10	Connect all cables at both ends, making certain that terminations are electrically and mechanically sound.	☐

Continued

E35 (CONTINUED)

THREE-PHASE DISTRIBUTION BOARD

OUTLINE OF MODIFIED INSTALLATION

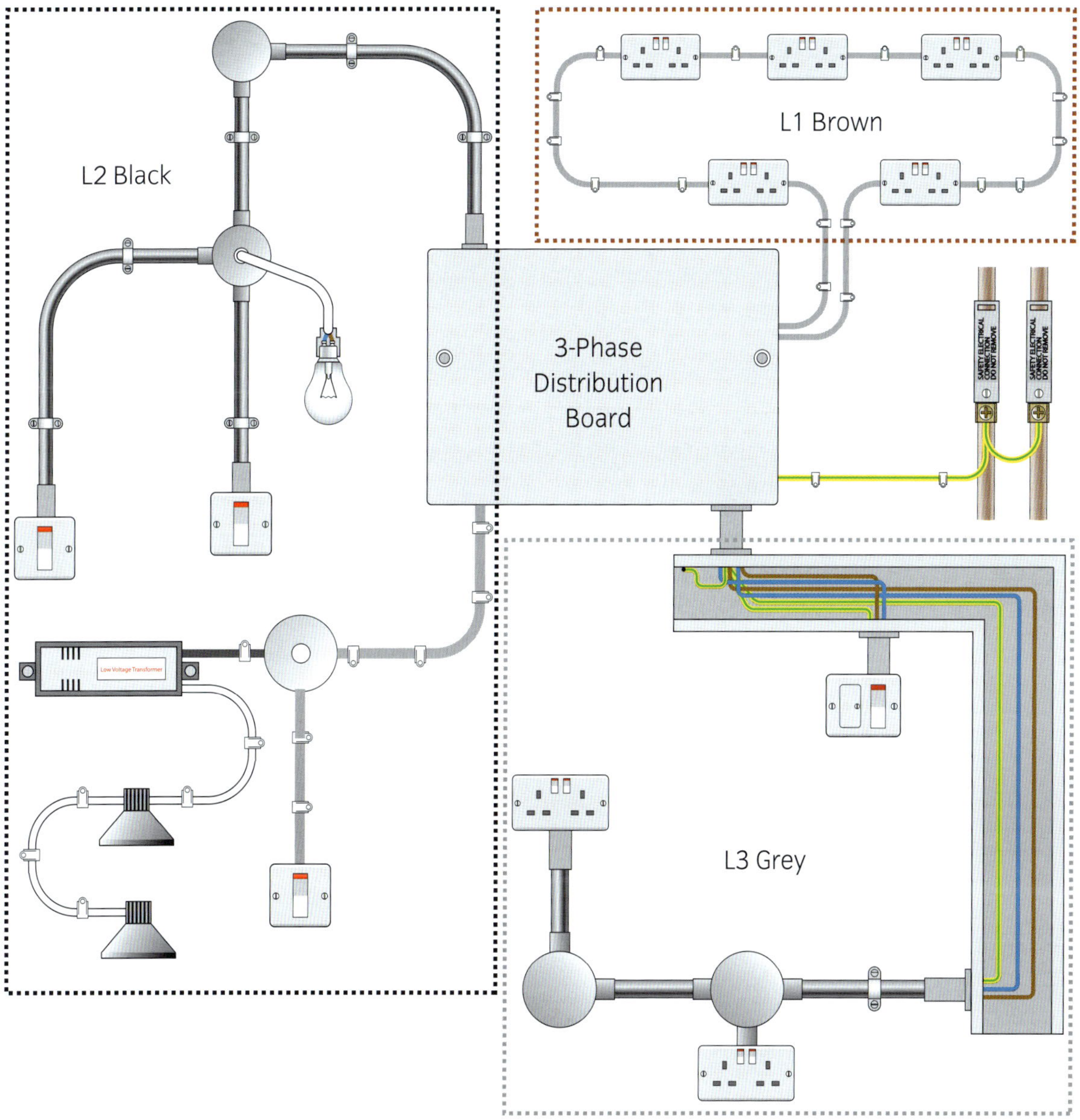

Date completed

Tutor/trainer's feedback/comments

E36

PERIODIC INSPECTION AND TESTING: THREE-PHASE SYSTEM

Perform periodic inspection and testing on a three-phase installation, and complete relevant documentation.

Note: this task requires you to complete an Electrical Installation Condition Report. The aim of this is to provide opportunity to gain greater experience in the techniques of inspection and testing, as well as familiarisation with the documents. You will need to approach each test using methods consistent with initial verification, as no previous test results are available.

If you have completed task E35, then your installation should look something like the illustration below. However, it might well be the case that for any number of reasons you have had to deviate from this specification. This should not prevent you from completing this task; simply complete the documentation for *your* installation.

It is not expected that you will complete this task in one single lesson period. You may wish to complete this task working with another learner.

Testing must be carried out in accordance with BS 7671, where the prescribed tests are detailed in the sequence in which they must be performed.

Details of how the tests are carried out are given in the IET publication *Guidance Note 3 – Inspection and Testing*. However, detailed instructions are also provided in the City & Guilds textbook *Level 3 NVQ Diploma in Electrotechnical Technology*, Units 307–308, Outcome 4.

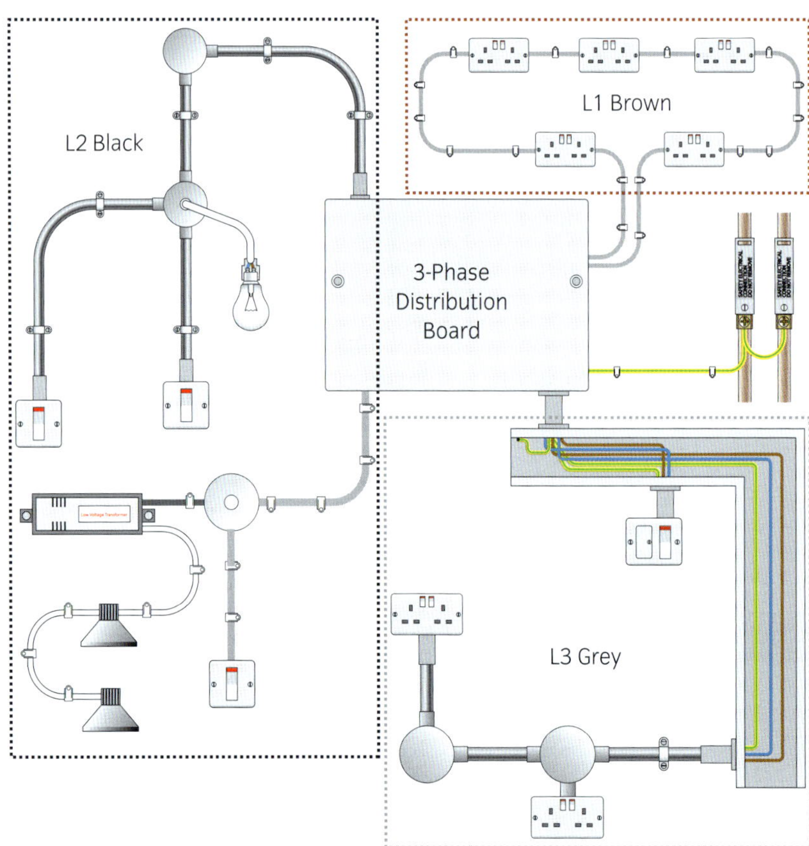

Continued

E36 (CONTINUED)

PERIODIC INSPECTION AND TESTING: THREE-PHASE SYSTEM

Complete the periodic inspection and testing forms provided in this task. The following notes relating to each part may help you. However, these are not intended to provide detailed instructions on how to perform the actual tests, such instructions may be found in *IET Guidance Note 3 – Inspection and Testing*, City & Guilds textbook *Level 3 NVQ Diploma in Electrotechnical Technology*, SmartScreen, or your course notes. Also, if you have previously completed task E34, you may find it helpful to refer to the results and notes made during that task.

TOP SECTION:

DB Ref No	This should be displayed on the cover of the distribution board.
Location	Enter the physical location of the distribution board.
Z_s at DB (Z_{db})	For a DB connected directly to the main incoming supply, this would be the Ze measurement. However, when measured at a sub-main board, the measurement is taken in the same manner, only it is referred to as Z_s at DB.
I_{pf} at DB	This test has to be performed live, using a suitable test instrument. However, the information may be obtained 'by enquiry' with the electricity supply company. Your tutor/trainer must be present whilst you are performing this test on a live circuit.
Correct supply polarity	This should be verified by live testing at the incoming supply using a suitable voltage test instrument.
Phase sequence	Using a test instrument that has a phase sequence test function, check that the phase sequence is correct at the incoming supply and – if relevant – at any other points in the installation where the three phases are present such as 3 phase motors, socket outlets, sub-mains distribution boards, etc. Your trainer must be present whilst you are performing this test.
Circuits vulnerable to damage	List any circuits in the installation that contain equipment/devices that may be damaged during testing. (These must be disconnected and bridged prior to testing.)
Details of test instruments	Make and serial number of test instrument(s) used for this testing.

Continued

E36 (CONTINUED)

PERIODIC INSPECTION AND TESTING: THREE-PHASE SYSTEM

COLUMNS 1–22:

Item	Complete all the tasks below	Candidate check ✓
1	The circuit number in the DB.	☐
2	Add a brief description (eg 'downstairs lighting').	☐
3	The BS EN number that defines the type of protective device for each circuit (eg BS EN60898). This information is found on the circuit breaker.	☐
4	The type of circuit breaker refers to the fault current at which the device will operate, breaking the circuit within 0.1 seconds. This is denoted by the letters B, C, or D. This information is found on the circuit breaker.	☐
5	The rating of the circuit breaker refers to the normal (no fault condition) maximum current that the device will pass in normal service.	☐
6	The breaking capacity is the maximum current that the device will withstand before catastrophic failure. Measured in kA. This information is found on the circuit breaker.	☐
7	This refers to the method of installation/containment of the cables, which has a bearing on the maximum current-carrying capacity. This information is found in BS 7671, Table 4A2.	☐
8	The cross-sectional area of the live conductors.	☐
9	The cross-sectional area of the cpc.	☐
10	This applies to ring-final circuits only. r_1 is the end-to-end resistance of the line conductor.	☐
11	This applies to ring-final circuits only. r_n is the end-to-end resistance of the neutral conductor.	☐
12	This applies to ring-final circuits only. r_2 is the end-to-end resistance of the cpc.	☐
13	This applies to all circuits. However, the test methods vary between lighting/radial circuits and ring-final circuits. For details on performing these tests, refer to *IET Guidance Note 3 – Inspection and Testing*.	☐
14	This applies to all circuits. This test is not required if $R_1 + R_2$ has been measured and recorded in column 13.	☐

Continued

E36 (CONTINUED)

PERIODIC INSPECTION AND TESTING: THREE-PHASE SYSTEM

Item	Complete all the tasks below	Candidate check ✓
15	**Important:** Before performing insulation resistance testing, the supply must be isolated, all items of equipment should be disconnected from sockets, light fittings, etc. All cpc and bonding conductors must be connected. All breakers must be set to ON. Test for two-way lighting circuits must be performed with the switches in both positions.	☐
16	The insulation resistance between the live and cpc conductors (line – cpc, and neutral – cpc).	☐
17	Once correct polarity has been confirmed, place a tick in this column.	☐
18	Having measured Z_{db} (that is; Z_s at DB) and $R_1 + R_2$, the total earth fault loop impedance $Z_s = Z_{db} + (R_1 + R_2)$. It is also possible to measure this value using an earth fault loop impedance tester where safe and suitable to do so. Note: The values obtained will vary depending on the external earthing method. Full details of these tests are given in *IET Guidance Note 3 – Inspection and Testing*.	☐
19	The breaking time of the RCD when tested at 1 x rated current.	☐
20	The breaking time of the RCD when tested at 5 x rated current. This test is not required for devices rated >30mA.	☐
21	This is a tick box confirming that the device operates correctly when the test button is pressed.	☐
22	Add comments on circuits, fault conditions (not relevant for an initial inspection), or any other relevant information.	☐

Continued

E36 (CONTINUED)

PERIODIC INSPECTION AND TESTING: THREE-PHASE SYSTEM

Tested by:

The certificate is only valid if this section is completed by the test engineer.

ELECTRICAL INSTALLATION CONDITION REPORT

SECTION A. DETAILS OF THE CLIENT / PERSON ORDERING THE REPORT

Name ....

Address ....

....

SECTION B. REASON FOR PRODUCING THIS REPORT ....

....

Date(s) on which inspection and testing was carried out ....

SECTION C. DETAILS OF THE INSTALLATION WHICH IS THE SUBJECT OF THIS REPORT

Occupier ....

Address ....

....

Description of premises (tick as appropriate)

Domestic ☐ Commercial ☐ Industrial ☐ Other (include brief description) ☐ ....

Estimated age of wiring system ....years

Evidence of additions / alterations Yes ☐ No ☐ Not apparent ☐ If yes, estimate age ....years

Installation records available? (Regulation 621.1) Yes ☐ No ☐ Date of last inspection .... (date)

SECTION D. EXTENT AND LIMITATIONS OF INSPECTION AND TESTING

Extent of the electrical installation covered by this report

....

....

Agreed limitations including the reasons (see Regulation 634.2) ....

....

Agreed with: ....

Operational limitations including the reasons (see page no.............) ....

....

The inspection and testing detailed in this report and accompanying schedules have been carried out in accordance with BS 7671: 2008 (IET Wiring Regulations) as amended to ....

It should be noted that cables concealed within trunking and conduits, under floors, in roof spaces, and generally within the fabric of the building or underground, have **not** been inspected unless specifically agreed between the client and inspector prior to the inspection.

SECTION E. SUMMARY OF THE CONDITION OF THE INSTALLATION

General condition of the installation (in terms of electrical safety) ....

....

....

Overall assessment of the installation in terms of its suitability for continued use

SATISFACTORY / UNSATISFACTORY* (Delete as appropriate)

*An unsatisfactory assessment indicates that dangerous (code C1) and/or potentially dangerous (code C2) conditions have been identified.

SECTION F. RECOMMENDATIONS

Where the overall assessment of the suitability of the installation for continued use above is stated as UNSATISFACTORY, I / we recommend that any observations classified as *'Danger present'* (code C1) or *'Potentially dangerous'* (code C2) are acted upon as a matter of urgency.

Investigation without delay is recommended for observations identified as *'further investigation required'*.

Observations classified as *'Improvement recommended'* (code C3) should be given due consideration.

Subject to the necessary remedial action being taken, I / we recommend that the installation is further inspected and tested by ....(date)

SECTION G. DECLARATION

I/We, being the person(s) responsible for the inspection and testing of the electrical installation (as indicated by my/our signatures below), particulars of which are described above, having exercised reasonable skill and care when carrying out the inspection and testing, hereby declare that the information in this report, including the observations and the attached schedules, provides an accurate assessment of the condition of the electrical installation, taking into account the stated extent and limitations in section D of this report.

Inspected and tested by:	**Report authorised for issue by:**
Name (Capitals)	Name (Capitals) ....
Signature	Signature ....
For/on behalf of	For/on behalf of ....
Position	Position ....
Address	Address ....
Date	Date ....

SECTION H. SCHEDULE(S)

...........schedule(s) of inspection and ....schedule(s) of test results are attached.

The attached schedule(s) are part of this document and this report is valid only when they are attached to it.

Continued

E36 (CONTINUED)

PERIODIC INSPECTION AND TESTING: THREE-PHASE SYSTEM

SECTION I. SUPPLY CHARACTERISTICS AND EARTHING ARRANGEMENTS

Earthing arrangements	Number and Type of Live Conductors		Nature of Supply Parameters	Supply Protective Device
TN-C ☐ TN-S ☐ TN-C-S ☐ TT ☐ IT ☐	a.c. ☐ 1-phase, 2-wire ☐ 2 phase, 3-wire ☐ 3 phase, 3-wire ☐ 3 phase, 4-wire ☐	d.c. ☐ 2-wire ☐ 3-wire ☐	Nominal voltage, U / U_0(1) .... V Nominal frequency, f(1) .... Hz Prospective fault current, I_{pf}(2) .... kA External loop impedance, Ze(2) .... Ω Note: (1) by enquiry (2) by enquiry or by measurement	BS (EN) .... Type .... Rated current .... A
	Confirmation of supply polarity ☐			

Other sources of supply (as detailed on attached schedule) ☐

SECTION J. PARTICULARS OF INSTALLATION REFERRED TO IN THE REPORT

Means of Earthing	**Details of Installation Earth Electrode** *(where applicable)*
Distributor's facility ☐	Type ....
Installation earth electrode ☐	Location ....
	Resistance to Earth .... Ω

Main Protective Conductors

Earthing conductor	Material	Csa ....mm^2	Connection / continuity verified ☐
Main protective bonding conductors	Material	Csa ....mm^2	Connection / continuity verified ☐
To incoming water service ☐	To incoming gas service ☐	To incoming oil service ☐	To structural steel ☐
To lightning protection ☐	To other incoming service(s) ☐ Specify		

Main Switch / Switch-Fuse / Circuit-Breaker / RCD

		If RCD main switch
Location	Current rating .... A	Rated residual operating current ($I_{\Delta n}$) .... mA
....	Fuse / device rating or setting .... A	Rated time delay .... ms
BS(EN)	Voltage rating V	Measured operating time(at $I_{\Delta n}$) .... ms
No of poles		

SECTION K. OBSERVATIONS

Referring to the attached schedules of inspection and test results, and subject to the limitations specified at the *Extent and limitations of inspection and testing* section

No remedial action is required ☐ The following observations are made ☐ (see below):

OBSERVATION(S)	CLASSIFICATION CODE	FURTHER INVESTIGATION REQUIRED (YES / NO)
....	....	....
....	....	....
....	....	....
....	....	....
....	....	....
....	....	....
....	....	....
....	....	....
....	....	....
....	....	....
....	....	....
....	....	....
....	....	....
....	....	....
....	....	....
....	....	....
....	....	....
....	....	....
....	....	....
....	....	....

One of the following codes, as appropriate, has been allocated to each of the observations made above to indicate to the person (s) responsible for the installation the degree of urgency for remedial action.

C1 – Danger present. Risk of injury. Immediate remedial action required

C2 – Potentially dangerous - urgent remedial action required

C3 – Improvement recommended

Continued

E36 (CONTINUED)

PERIODIC INSPECTION AND TESTING: THREE-PHASE SYSTEM

CONDITION REPORT INSPECTION SCHEDULE FOR DOMESTIC AND SIMILAR PREMISES WITH UP TO 100 A SUPPLY

Note: This form is suitable for many types of smaller installation not exclusively domestic.

OUTCOMES	*Acceptable condition* ✓	*Unacceptable condition* *State* **C1** *or* **C2**	*Improvement recommended* *State* **C3**	*Not verified* **N/V**	*Limitation* **LIM**	*Not applicable* **N/A**

ITEM NO	DESCRIPTION	OUTCOME (*Use codes above. Provide additional comment where appropriate. C1, C2 and C3 coded items to be recorded in Section K of the Condition Report*)	*Further investigation required?* (**Y** or **N**)
1.0	**DISTRIBUTOR'S / SUPPLY INTAKE EQUIPMENT**		
1.1	Service cable condition		
1.2	Condition of service head		
1.3	Condition of tails - Distributor		
1.4	Condition of tails - Consumer		
1.5	Condition of metering equipment		
1.6	Condition of isolator (where present)		
2.0	**PRESENCE OF ADEQUATE ARRANGEMENTS FOR OTHER SOURCES SUCH AS MICROGENERATORS (551.6; 551.7)**		
3.0	**EARTHING / BONDING ARRANGEMENTS (411.3; Chap 54)**		
3.1	Presence and condition of distributor's earthing arrangement (542.1.2.1; 542.1.2.2)		
3.2	Presence and condition of earth electrode connection where applicable (542.1.2.3)		
3.3	Provision of earthing / bonding labels at all appropriate locations (514.11)		
3.4	Confirmation of earthing conductor size (542.3; 543.1.1)		
3.5	Accessibility and condition of earthing conductor at MET (543.3.2)		
3.6	Confirmation of main protective bonding conductor sizes (544.1)		
3.7	Condition and accessibility of main protective bonding conductor connections (543.3.2; 544.1.2)		
3.8	Accessibility and condition of all protective bonding connections (543.3.2)		
4.0	**CONSUMER UNIT(S) / DISTRIBUTION BOARD(S)**		
4.1	Adequacy of working space / accessibility to consumer unit / distribution board (132.12; 513.1)		
4.2	Security of fixing (134.1.1)		
4.3	Condition of enclosure(s) in terms of IP rating etc (416.2)		
4.4	Condition of enclosure(s) in terms of fire rating etc (526.5)		
4.5	Enclosure not damaged/deteriorated so as to impair safety (621.2(iii))		
4.6	Presence of main linked switch (as required by 537.1.4)		
4.7	Operation of main switch (functional check) (612.13.2)		
4.8	Manual operation of circuit-breakers and RCDs to prove disconnection (612.13.2)		
4.9	Correct identification of circuit details and protective devices (514.8.1; 514.9.1)		
4.10	Presence of RCD quarterly test notice at or near consumer unit / distribution board (514.12.2)		
4.11	Presence of non-standard (mixed) cable colour warning notice at or near consumer unit / distribution board (514.14)		
4.12	Presence of alternative supply warning notice at or near consumer unit / distribution board (514.15)		
4.13	Presence of other required labelling (please specify) (Section 514)		
4.14	Examination of protective device(s) and base(s); correct type and rating (no signs of unacceptable thermal damage, arcing or overheating) (421.1.3)		
4.15	Single-pole protective devices in line conductor only (132.14.1; 530.3.2)		
4.16	Protection against mechanical damage where cables enter consumer unit / distribution board (522.8.1; 522.8.11)		
4.17	Protection against electromagnetic effects where cables enter consumer unit / distribution board / enclosures (521.5.1)		
4.18	RCD(s) provided for fault protection – includes RCBOs (411.4.9; 411.5.2; 531.2)		
4.19	RCD(s) provided for additional protection - includes RCBOs (411.3.3; 415.1)		

Continued

E36 (CONTINUED)

PERIODIC INSPECTION AND TESTING: THREE-PHASE SYSTEM

OUTCOMES	Acceptable condition	✓	Unacceptable condition	State C1 or C2	Improvement recommended	State C3	Not verified	N/V	Limitation	LIM	Not applicable	N/A

ITEM NO	DESCRIPTION	OUTCOME (Use codes above. Provide additional comment where appropriate. C1, C2 and C3 coded items to be recorded in Section K of the Condition Report)	Further investigation required? (Y or N)
5.0	**FINAL CIRCUITS**		
5.1	Identification of conductors (514.3.1)		
5.2	Cables correctly supported throughout their run (522.8.5)		
5.3	Condition of insulation of live parts (416.1)		
5.4	Non-sheathed cables protected by enclosure in conduit, ducting or trunking (521.10.1)		
	▪ To include the integrity of conduit and trunking systems (metallic and plastic)		
5.5	Adequacy of cables for current-carrying capacity with regard for the type and nature of installation (Section 523)		
5.6	Coordination between conductors and overload protective devices (433.1; 533.2.1)		
5.7	Adequacy of protective devices: type and rated current for fault protection (411.3)		
5.8	Presence and adequacy of circuit protective conductors (411.3.1.1; 543.1)		
5.9	Wiring system(s) appropriate for the type and nature of the installation and external influences (Section 522)		
5.10	Concealed cables installed in prescribed zones (see Section D. *Extent and limitations*) (522.6.101)		
5.11	Concealed cables incorporating earthed armour or sheath, or run within earthed wiring system, or otherwise protected against mechanical damage from nails, screws and the like (see Section D. *Extent and limitations*) (522.6.101; 522.6.103)		
5.12	Provision of additional protection by RCD not exceeding 30 mA:		
	▪ for all socket-outlets of rating 20 A or less provided for use by ordinary persons unless an exception is permitted (411.3.3)		
	▪ for supply to mobile equipment not exceeding 32 A rating for use outdoors (411.3.3)		
	▪ for cables concealed in walls or partitions (522.6.102; 522.6.103)		
5.13	Provision of fire barriers, sealing arrangements and protection against thermal effects (Section 527)		
5.14	Band II cables segregated / separated from Band I cables (528.1)		
5.15	Cables segregated / separated from communications cabling (528.2)		
5.16	Cables segregated / separated from non-electrical services (528.3)		
5.17	Termination of cables at enclosures – indicate extent of sampling in Section D of the report (Section 526)		
	▪ Connections soundly made and under no undue strain (526.6)		
	▪ No basic insulation of a conductor visible outside enclosure (526.98)		
	▪ Connections of live conductors adequately enclosed (526.5)		
	▪ Adequately connected at point of entry to enclosure (glands, bushes etc.) (522.8.5)		
5.18	Condition of accessories including socket-outlets, switches and joint boxes (621.2 (iii))		
5.19	Suitability of accessories for external influences (512.2)		
6.0	**LOCATION(S) CONTAINING A BATH OR SHOWER**		
6.1	Additional protection for all low voltage (LV) circuits by RCD not exceeding 30 mA (701.411.3.3)		
6.2	Where used as a protective measure, requirements for SELV or PELV met (701.414.4.5)		
6.3	Shaver sockets comply with BS EN 61558-2-5 formally BS 3535 (701.512.3)		
6.4	Presence of supplementary bonding conductors, unless not required by BS 7671:2008 (701.415.2)		
6.5	Low voltage (e.g. 230 volt) socket-outlets sited at least 3 m from zone 1 (701.512.3)		
6.6	Suitability of equipment for external influences for installed location in terms of IP rating (701.512.2)		
6.7	Suitability of equipment for installation in a particular zone (701.512.3)		
6.8	Suitability of current-using equipment for particular position within the location (701.55)		
7.0	**OTHER PART 7 SPECIAL INSTALLATIONS OR LOCATIONS**		
7.1	List all other special installations or locations present, if any. (Record separately the results of particular inspections applied.)		

Inspected by:
Name (Capitals) .... **Signature** .... **Date** ....

Continued

E36 (CONTINUED)

PERIODIC INSPECTION AND TESTING: THREE-PHASE SYSTEM

Form 4 | Form No: ..../4

GENERIC SCHEDULE OF TEST RESULTS

DB reference no ....
Location ....
Zs at DB (Ω) ....
I_{pf} at DB (kA) ....
Correct supply polarity confirmed ☐
Phase sequence confirmed (where appropriate) ☐

Details of circuits and/or installed equipment vulnerable to damage when testing ....
....
....
....
....

Details of test instruments used (state serial and/or asset numbers)
Continuity ....
Insulation resistance ....
Earth fault loop impedance ....
RCD ....
Earth electrode resistance ....

Tested by:
Name (Capitals) ....
Signature .... **Date** ....

Circuit details									**Test results**												
		Overcurrent device				Conductor details			Ring final circuit continuity (Ω)			Continuity (Ω) ($R_1 + R_2$) or R_2		Insulation Resistance (MΩ)		Polarity	Z_s (Ω)	RCD (ms)			Remarks (continue on a separate sheet if necessary)
Circuit number 1	Circuit Description 2	BS (EN) 3	type 4	rating (A) 5	breaking capacity (kA) 6	Reference Method 7	Live (mm²) 8	cpc (mm²) 9	r_1 (line) 10	r_n (neutral) 11	r_2 (cpc) 12	($R_1 + R_2$) * 13	R_2 14	Live-Live 15	Live-E 16	17	18	@ $I_{\Delta n}$ 19	@ $5I_{\Delta n}$ 20	Test button operation 21	22

* Where there are no spurs connected to a ring final circuit this value is also the ($R_1 + R_2$) of the circuit.

Continued

E36 (CONTINUED)

PERIODIC INSPECTION AND TESTING: THREE-PHASE SYSTEM

Date completed

Tutor/trainer's feedback/comments

E37

THREE-PHASE MOTOR INSTALLATION AND TESTING

INSTALL, CONNECT AND TEST A THREE-PHASE MOTOR CIRCUIT:

The aim of this task is to practice connection and testing of a three-phase circuit. Use is made of the existing three-phase installation that you have completed in task E35, as well as the tray assembly completed in tasks E17/E18. However, your tutor/trainer may wish to modify this task to suit the available equipment, and the manner in which you performed the previous tasks.

If you are following on from the previous tasks, assume that the customer has requested the installation of an extraction fan which involves the installation of an SWA supply from the distribution board to a three-phase switch disconnector, single-core cables contained in PVC conduit to the DOL starter, and a connection contained in flexible conduit to a three-phase motor. The installation is illustrated below.

Important: before commencing this task, ensure that the supply is isolated and locked off. Prove this with your tutor/trainer before proceeding.

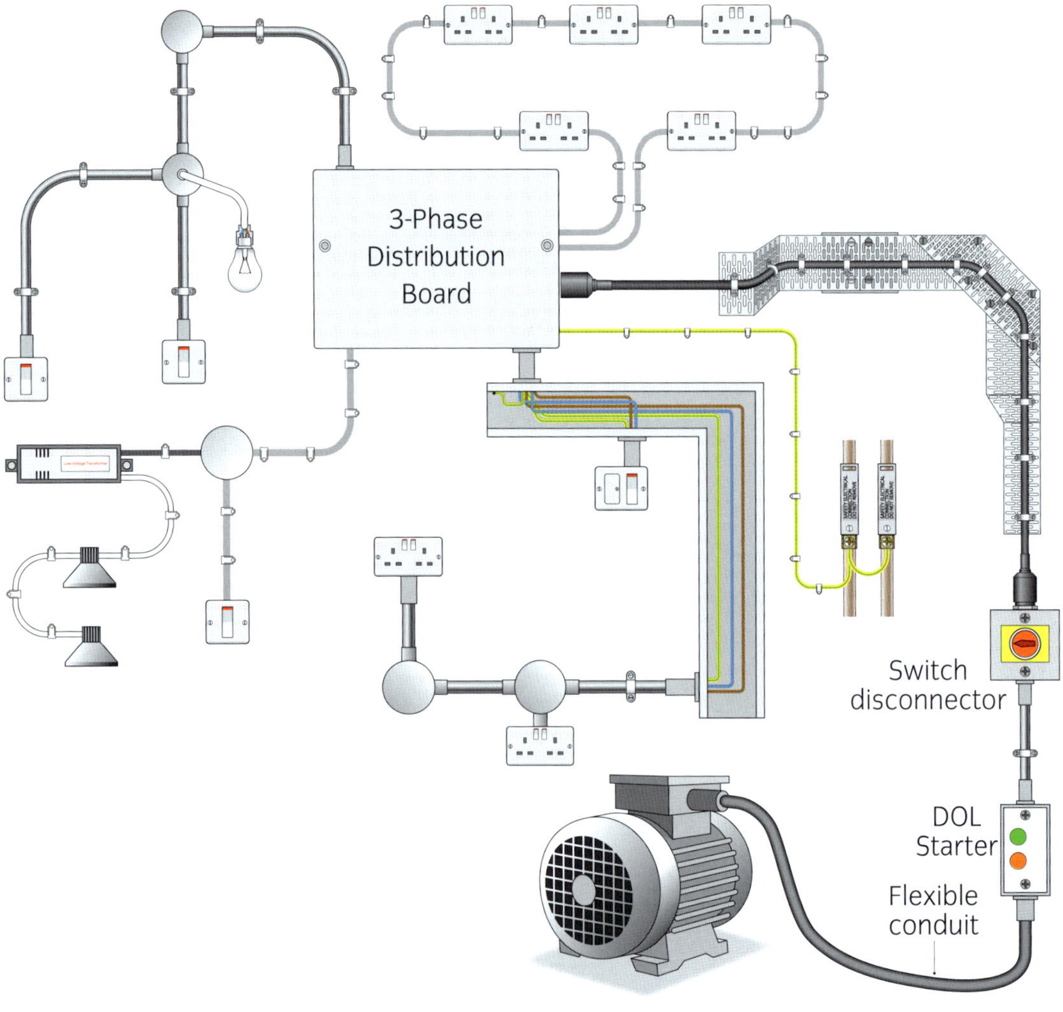

Continued

E37 (CONTINUED)

THREE-PHASE MOTOR INSTALLATION AND TESTING

INSTALLATION:

Item	Complete all the tasks below	Candidate check ✓
1	Remove the SWA cable from the tray produced in tasks E17/E18.	☐
2	Fix the tray to the surface as illustrated, or to suit your particular installation. Also fix the DOL starter and three-phase switch disconnector.	☐
3	Measure and cut a length of $2.5mm^2$ four-core SWA cable to connect between the distribution board and the three-phase switch disconnector.	☐
4	Fix the SWA cable to the tray using cable ties.	☐
5	Remove suitable lengths of outer sleeve and insulation from the SWA cable ends, leaving sufficient lengths of conductor to complete the connections in the distribution board and the switch disconnector.	☐
6	Connect the SWA cable to the distribution board and switch disconnector using suitable glands and banjos. Use a suitably sized conductor to connect the armouring to the earth terminal using a banjo on the SWA gland. Fit shrouds over each connection.	☐
7	Connect the four line connectors in the distribution board to a suitably rated three-pole circuit breaker, and neutral. Ensure that the correct phase sequence is observed.	☐
8	Connect the four line connectors and cpc in the switch disconnector.	☐
9	Measure and cut a length of 20mm PVC conduit to connect between the switch disconnector and the DOL starter. Ream and de-burr as required, and fit the conduit using manufacturer-approved methods.	☐
10	Measure and cut lengths of $2.5mm^2$ conductor for the L1, L2, L3, neutral and cpc connections between the switch disconnector and the DOL starter, allowing for connection inside the enclosures. Pass these through the conduit and complete all connections at both ends, making sure that the correct phase sequence is maintained.	☐

Continued

E37 (CONTINUED)

THREE-PHASE MOTOR INSTALLATION AND TESTING

Item	Complete all the tasks below	Candidate check ✓
11	Measure and cut a length of 20mm flexible conduit to fit between the DOL starter and the motor connection point.	☐
12	Measure and cut lengths of 2.5mm2 conductor for the L1, L2, L3, and cpc connections between the DOL starter and the motor. Pass these through the conduit and then fix the conduit to the DOL starter and motor using manufacturer-approved methods. (Depending on your installation, you may need to support the conduit using saddles).	☐
13	Connect the conductors in the DOL starter, making sure that the correct phase sequence is maintained. **Do not connect the cables in the motor** (the circuit must first be tested with the motor disconnected); however, make the conductors safe by terminating them in terminal blocks and wrapping them in insulation tape, or by termination in a junction box.	☐

TESTING:

Add the new installation to the generic schedule of test results in task E36, performing all necessary tests. You may find it helpful to refer to the guide notes to completing the test schedule in task E36.

Note 1: before commencing testing, make sure that the starter and switch disconnector are set to the ON position, and that all cpc connections are in place.

Note 2: $R_1 + R_2$ measurement is performed in a similar manner to a single-phase circuit, except you connect L1/L2/L3 to the cpc at the DB, and you then test between each line and the cpc.

FUNCTIONAL TESTING:

This part of the task may best be performed as a group, under the direction of your tutor/trainer.

It is not actually necessary to connect a motor to prove that the circuit is functioning (your testing should have proven that). However, if you do have access to a three-phase motor that you can connect to the circuit, you may complete the following functional tests.

Note: the motor shaft should not be physically connected to any device that is direction sensitive during these tests.

Continued

E37 (CONTINUED)

THREE-PHASE MOTOR INSTALLATION AND TESTING

Item	Complete all the tasks below	Candidate check ✓
1	Isolate and lock off the supply using the three-phase switch disconnector, and fit an appropriate warning sign.	☐
2	Reinstate the supply to the main DB.	☐
3	Use an approved test instrument to confirm that all line conductors at the motor are dead **(your tutor/trainer must be present during this test)**.	☐
4	Connect the line and cpc conductors in the motor. Replace the motor cover.	☐
5	Turn the supply on at the switch disconnector and operate the DOL starter switch. Note the direction of rotation of the motor.	☐
6	Switch off the motor, and repeat steps 1 and 3 above.	☐
7	Change the phase connections in the motor by reversing L1 and L3.	☐
8	Turn the supply on at the switch disconnector and operate the DOL starter switch. Note the direction of rotation of the motor.	☐
9	Switch off the motor, and repeat steps 1 and 3 above.	☐
10	Change the phase connections in the motor by reversing L2 and L3.	☐
11	Turn the supply on at the switch disconnector and operate the DOL starter switch. Note the direction of rotation of the motor.	☐

Date completed

Tutor/trainer's feedback/comments

E38
FAULT LOCATION

DIAGNOSE AND RECTIFY FAULTS IN ELECTRICAL INSTALLATIONS:

The aim of this task is to provide opportunity to practise a logical approach to fault location in electrical circuits. If you have completed the installation in combined tasks E25 to E32, you will now use this installation as a fault location practise rig. If you have also completed task E35, then you may use the three-phase installation as your practise rig. Alternatively, your centre may have other installations that they wish to use for this exercise.

Ideally, you will work in pairs.

Important: before commencing this task, full isolation must be undertaken.

Item	Complete all the tasks below	Candidate check ✓
1	Isolate and lock off the incoming supply and fit an appropriate warning sign.	☐
2	Use an approved test instrument to confirm that all line conductors at the incoming supply are dead **(your tutor/trainer must be present during this test)**.	☐

FAULT INSERTION:

Item	Complete all the tasks below	Candidate check ✓
1	Referring to a layout diagram of your installation, identify three realistic fault conditions that could be inserted into the installation. The diagram for the three-phase combined task in this manual is given below, along with some fault examples.	☐
2	Write down your three fault conditions in the space below and confirm with your tutor/trainer whether they are realistic. Your tutor/trainer may be able to assist you by providing you with defective components to insert into the installation.	☐
3	Insert the three faults into the installation of another learner without them being able to observe (at this time they might be inserting three faults into your installation).	☐
4	On a separate piece of paper, write down a brief description of the fault symptoms experienced by the 'customer' for the three faults that you have inserted, and pass this to the learner, who is to diagnose and rectify the faults.	☐

Continued

E38 (CONTINUED)

FAULT LOCATION

Fault 1:

Fault 2:

Fault 3:

TESTING AND FAULT LOCATION:

Once another learner has inserted three faults into your installation, using the list of fault symptoms experienced by the 'customer' and the results from your testing; diagnose and rectify the faults.

Item	Complete all the tasks below	Candidate check ✓
1	Complete the Generic Schedule of Test Results sheet provided in this task. Each fault should become apparent as you work through the tests.	☐
2	As you identify each faulty circuit, use appropriate test instruments, and a logical test procedure, to locate the fault(s). Correct the fault and re-test the relevant circuit.	☐
3	Write down each fault in the spaces below.	☐
4	When all faults have been rectified, confirm with your colleagues that you have corrected the faults that they introduced.	☐
5	As a group, discuss the methods and applied tests that you used to diagnose each fault.	☐

Diagnosed fault 1:

Diagnosed fault 2:

Diagnosed fault 3:

Continued

E38 (CONTINUED)

FAULT LOCATION

THREE-PHASE INSTALLATION SPECIFIED IN TASKS E25 TO E32, AND TASK E35

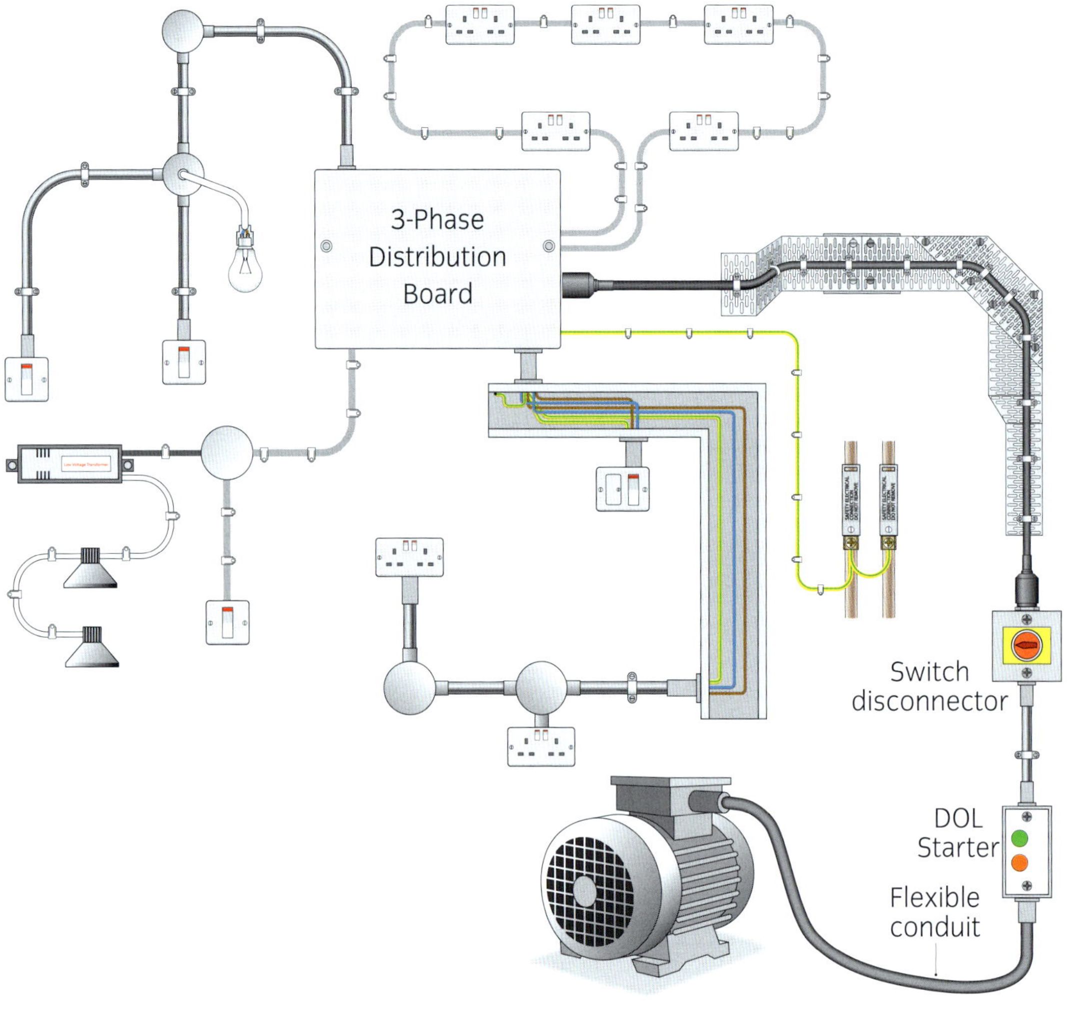

Continued

E38 (CONTINUED)
FAULT LOCATION

FAULT EXAMPLES:

Note: the faults listed below are for example only.

1. Remove one cpc conductor from one of the socket-outlets in the ring final circuit.
2. Reverse the polarity of the line and neutral conductors in any one of the socket outlets.
3. Simulate a broken conductor by removing the neutral conductor from the PVC conduit in the two-way lighting circuit, cut the conductor, and then re-insert the conductor.
4. Swop the connections between the 20A circuit breaker on the radial circuit and the 32A circuit breaker on the ring final circuit. (If the cables are not long enough, swop the CBs around.)
5. Insert a 10kΩ resistor between the line and cpc terminals in one socket outlet in the radial circuit.
6. Depending on availability of defective components, you could insert defective components such as:
    a. an RCD that has a trip time well outside of accepted times
    b. a circuit breaker that has an open circuit contact
    c. a switch with permanently closed contacts in one pole, etc.

Continued

Form 4

Form No: ..../4

GENERIC SCHEDULE OF TEST RESULTS

DB reference no ....
Location ....
Zs at DB (Ω) ....
I_{pf} at DB (kA) ....
Correct supply polarity confirmed ☐
Phase sequence confirmed (where appropriate) ☐

Details of circuits and/or installed equipment vulnerable to damage when testing ....

Details of test instruments used (state serial and/or asset numbers)
Continuity ....
Insulation resistance ....
Earth fault loop impedance ....
RCD ....
Earth electrode resistance ....

Tested by:
Name (Capitals) ....
Signature .... **Date** ....

Circuit details									Test results												
		Overcurrent device				Conductor details			Ring final circuit continuity (Ω)			Continuity (Ω) ($R_1 + R_2$) or R_2		Insulation Resistance (MΩ)		Polarity	Z_s (Ω)	RCD (ms)			Remarks (continue on a separate sheet if necessary)
Circuit number 1	Circuit Description 2	BS (EN) 3	type 4	rating (A) 5	breaking capacity (kA) 6	Reference Method 7	Live (mm²) 8	cpc (mm²) 9	r_1 (line) 10	r_n (neutral) 11	r_2 (cpc) 12	($R_1 + R_2$) * 13	R_2 14	Live-Live 15	Live-E 16	17	18	@ $I_{\Delta n}$ 19	@ $5I_{\Delta n}$ 20	Test button operation 21	22

* Where there are no spurs connected to a ring final circuit this value is also the ($R_1 + R_2$) of the circuit.

Continued

E38 (CONTINUED)

FAULT LOCATION

Date completed

Tutor/trainer's feedback/comments

E39

DATA CABLE AND SOCKET CONNECTION

Connect two network sockets using TIA wiring standard T568B.

Note: whether your tutor/trainer provides you with CAT5e or CAT6e cable will not make any significant difference to this practical activity. However, because of the improved performance and the falling costs of hardware, CAT6e is currently used in any new data network installation.

Item	Complete all the tasks below	Candidate check ✓
1	Fix the two socket backboxes and PVC trunking to a surface, as illustrated.	☐
2	Measure and cut a length of CAT5e/CAT6e data cable, leaving sufficient for termination.	☐
3	Remove a suitable length of cable sheath at both ends using a rotary stripper.	☐
4	Terminate the conductors on each socket using the colour convention in TIA wiring standard T568B, using a punch down tool.	☐
5	Fix each faceplate, ensuring correct dressing of conductors inside each backbox, to TIA standards.	☐
6	Using a suitable data cable test instrument, test the circuit for polarity and continuity.	☐
7	Dismantle the installation after inspection by your tutor/trainer.	☐

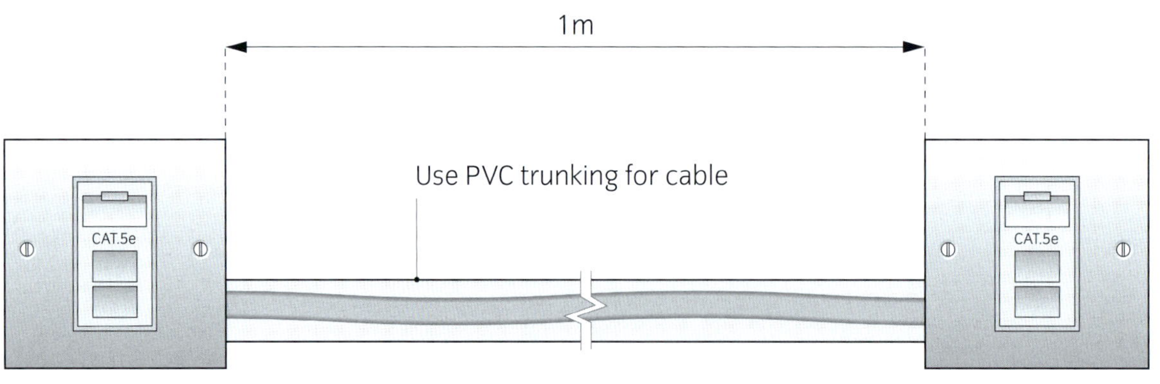

Date completed

Tutor/trainer's feedback/comments

SCIENCE TASKS

The following tasks relate to science components in the course. They are designed to be standalone and may be performed in any order. These tasks should be inserted at appropriate points into the scheme of work to help consolidate the relevant theory.

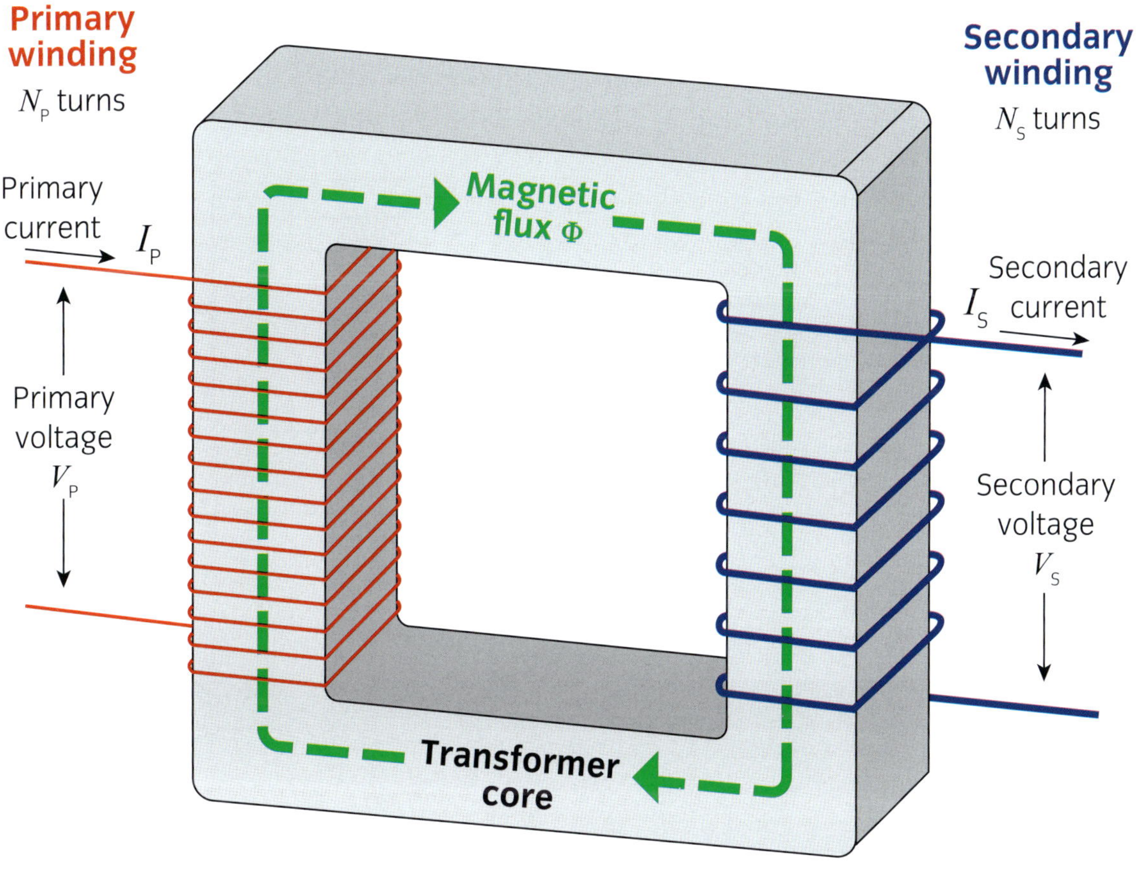

E40

SOLDER CONNECTIONS

This work will be used in tasks E41 and E42.

Construct a simple extra-low-voltage electrical circuit, using soldered terminals and connections.

Note to tutor/trainer: The resistor values have not been specified, as this will depend on available components at your centre. However, task E41 relies on the following conditions:

R1 = R2 R3 = R4 R1/R2 are not equal to R3/R4

In order to obtain meaningful current readings in task E41 (assuming that a dc supply of 9–12Vdc is to be used), the resistor values should be in the order of 10Ω to 47Ω; 1W minimum rating.

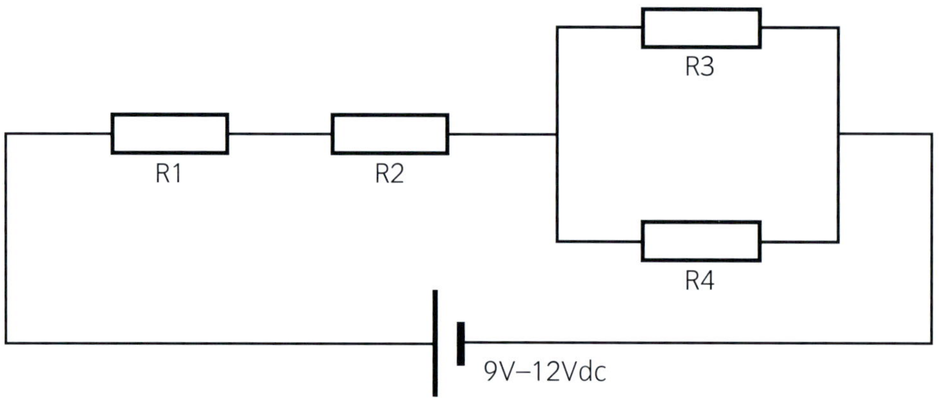

In the table below, write down the values provided by your tutor/trainer.

Resistor	Value
R1	
R2	
R3	
R4	

Continued

E40 (CONTINUED)

SOLDER CONNECTIONS

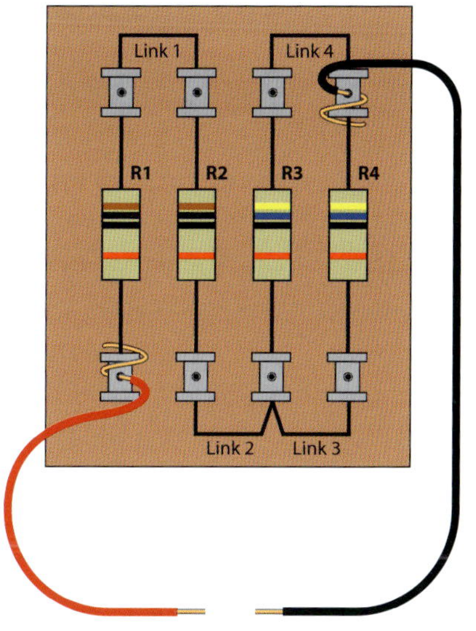

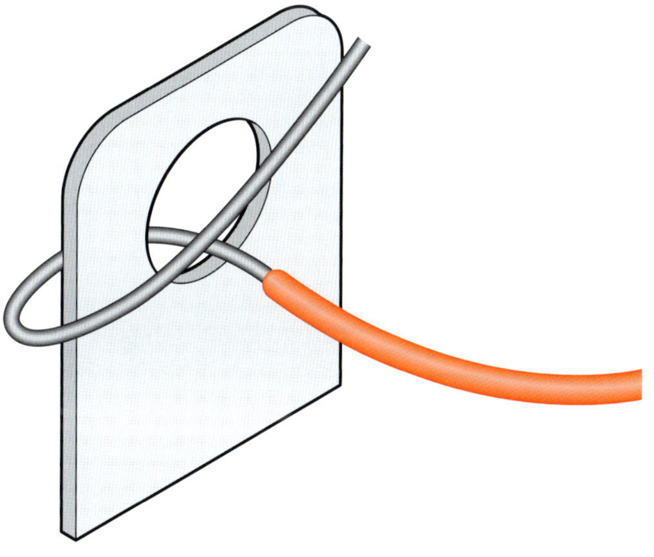

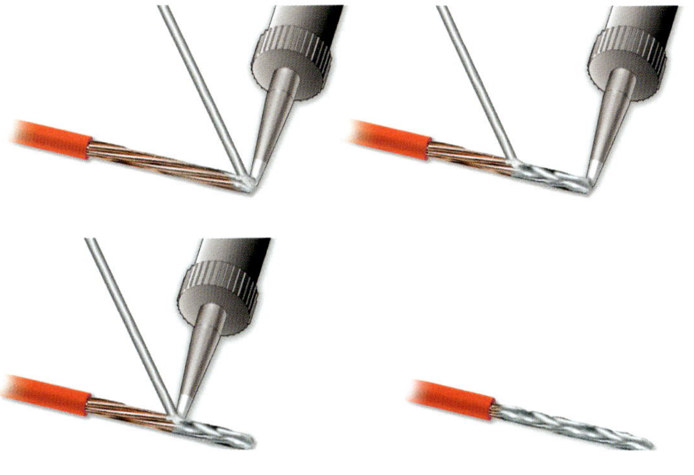

Continued

E40 (CONTINUED)
SOLDER CONNECTIONS

Item	Complete all the tasks below	Candidate check ✓
1	Identify the value of each of the resistors. You may do this by reading the colour code, or by measurement using a digital multimeter.	☐
2	Mount resistors R1 and R2 onto the tag board in the positions shown in the layout diagram, wrapping each wire once around the tag using long-nosed pliers. Cut off excess wire ends.	☐
3	Cut two lengths of uninsulated wire in the order of 22 SWG to make links 1 and 2. Using long-nosed pliers, form the links to have square corners and suitable width to fit onto the tag board. Fix the two links to the tag board as shown in the layout diagram.	☐
4	Cut a length of red insulated wire to connect the board to the dc power supply. Strip 10mm of insulation from one end.	☐
5	Tin the wire using a soldering iron, as illustrated.	☐
6	Wrap the tinned end of the wire around the tag in the position shown in the layout diagram.	☐
7	Solder the terminals at both ends of R1 and R2. Do not solder the other end of link 2 at this stage.	☐
8	Mount resistors R3 and R4 onto the tag board in the positions shown in the layout diagram, wrapping each wire once around the tag using long-nosed pliers. Cut off excess wire ends.	☐
9	Cut two lengths of uninsulated wire in the order of 22 SWG to make links 3 and 4. Using long-nosed pliers, form the links to have square corners and suitable width to fit onto the tag board. Fix the two links to the tag board as shown in the layout diagram.	☐
10	Solder the terminals at both ends of R3.	☐
11	Cut a length of black insulated wire to connect the board to the dc power supply. Strip 10mm of insulation from one end.	☐
12	Tin the wire using a soldering iron, as illustrated.	☐
13	Wrap the tinned end of the wire around the tag in the position shown in the layout diagram.	☐
14	Solder the terminals at both ends of R4.	☐
15	Strip sufficient insulation from the ends of the red and black wires to allow connection to the dc supply.	☐

Date completed

Tutor/trainer's feedback/comments

E41

INTRODUCTION TO TEST INSTRUMENTS

Perform simple tests using common electrical test instruments.

Note: these instructions provide an outline of common tests that you might perform when testing an electrical/electronic circuit. However, the actual selection and setting up of the test instrument(s) will be determined by the specific make and type of test instrument(s) used at your centre.

MEASURING RESISTANCE:

To measure the resistance of a component or cable:

- the component or cable must be dead (no power connected)
- the test instrument must be set to the resistance/Ohms range; if the instrument is not auto-ranging, then a resistance range that relates to the expected resistance of the component/cable must also be selected
- the test leads are connected to the terminal/conductor at either end of the component/cable (see diagram).

Your tutor/trainer will provide you with the electrical items listed in table 1. Use a test meter to measure the resistance value of each, recording the reading in the table.

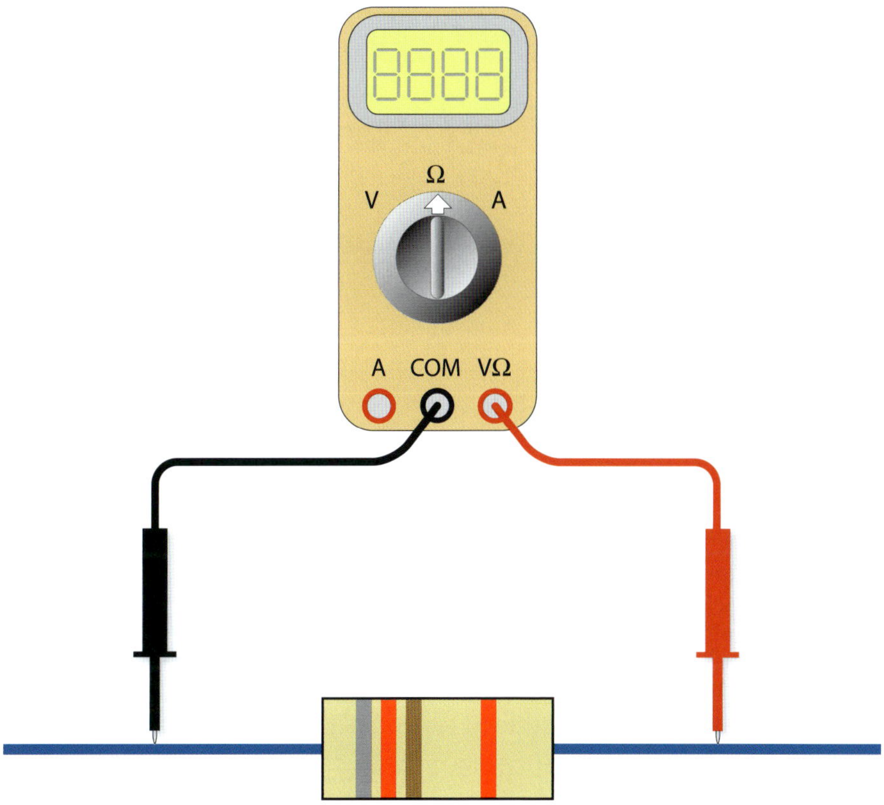

Continued

E41 (CONTINUED)

INTRODUCTION TO TEST INSTRUMENTS

In the table below, write down the values provided by your tutor/trainer.

Table 1

Component	Measured resistance
Resistor 1	
Resistor 2	
Resistor 3	
Tungsten lamp	
1m length of 0.75mm flex	
100m reel of 2.5mm^2 flat T&E cable (measure the end-to-end resistance of the brown conductor)	
100m reel of 2.5mm^2 flat T&E cable (measure the end-to-end resistance of the cpc)	

MEASURING VOLTAGE:

To measure the voltage across a component or electrical terminals:

- the circuit must be energised
- you must always first inspect the test instrument and leads to make sure that it is safe to use (ie, no damage to casing, no cracks/breaks in insulation)
- the test instrument must be set to the voltage range; if the instrument is not auto-ranging, then a high voltage range is normally selected to begin with then, if the voltage is found to be much lower, the range can be reduced to obtain a more accurate reading
- the test leads are connected to the terminal/conductor at either end of the component, or to each of the terminals.

Note: when testing anything other than extra low voltage circuits, you should connect the negative/neutral lead first, followed by the positive/live connector.

Continued

E41 (CONTINUED)

INTRODUCTION TO TEST INSTRUMENTS

Construct the circuit below using a 5A connector block. Make certain that the light emitting diode (LED) is connected the correct may around (the cathode (k) is usually indicated by a flat edge on the body of the LED).

Energise your circuit either using a dc power supply set to 12Vdc output, or a 9Vdc battery.

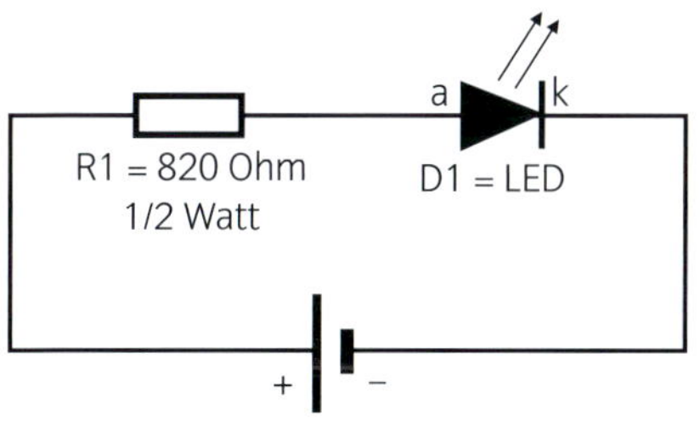

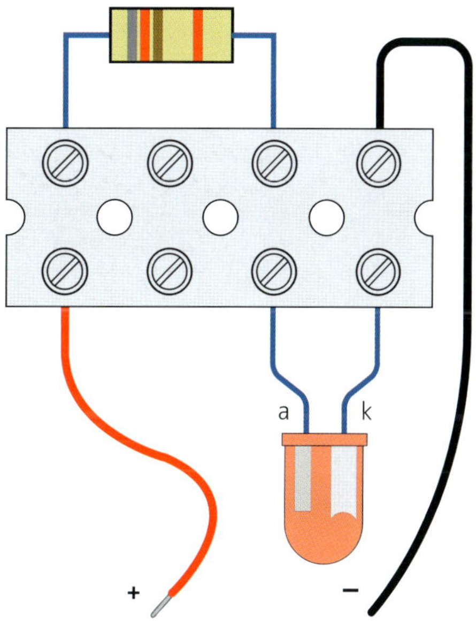

Use a test meter to measure the voltage at each of the points listed below, recording the readings.

Test point	Measured dc voltage
Across the dc power supply terminals	
Across resistor R1	
Across diode D1	

Continued

MEASURING LOW CURRENTS:

To measure the current in an electronic circuit:

- you must always first inspect the test instrument and leads to make sure that it is safe to use (ie, no damage to casing, no cracks/breaks in insulation)
- the test instrument must be set to the ac or dc current range. If the instrument is not auto-ranging, a high current range is normally selected to begin with then, if the current is found to be much lower, the range can be reduced to obtain a more accurate reading
- make sure that the circuit is de-energised; remove the power input lead, and connect the test instrument in series with the current flow (see diagram).

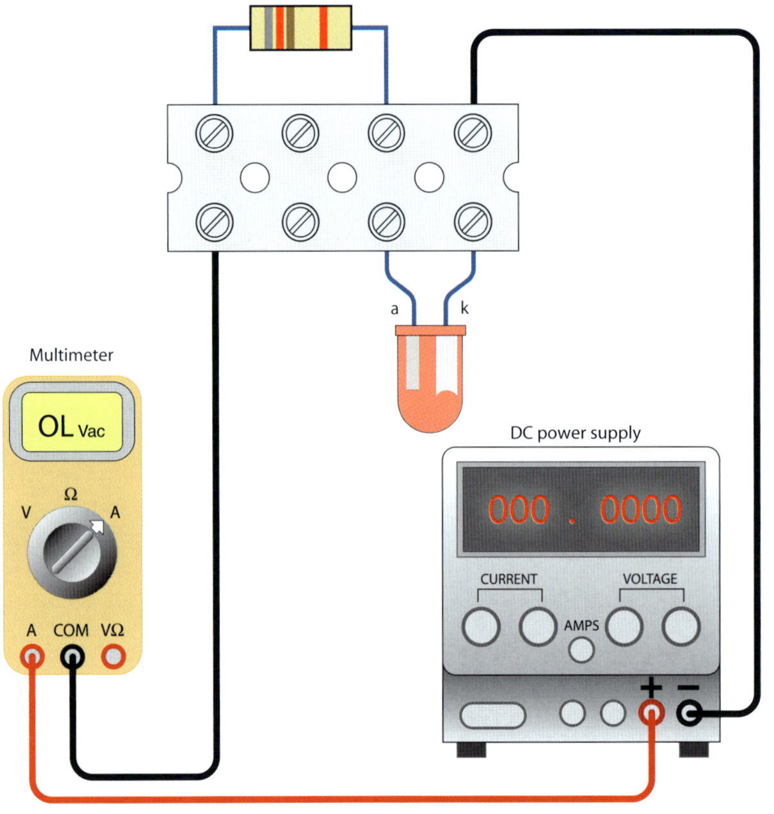

Use a test meter to measure the current taken by the LED circuit. Record the reading below.

Circuit current = ______________ A

Date completed

Tutor/trainer's feedback/comments

E42
RESISTIVE CIRCUIT THEORY

Use the tag board constructed in task E40.

INVESTIGATE V I R RELATIONSHIPS IN A RESISTIVE CIRCUIT:

Using the resistive circuit constructed in task E40, you will use a multimeter to prove theoretical calculations relating to this circuit.

Note: the term R1//R2 used below means R1 in parallel with R2.

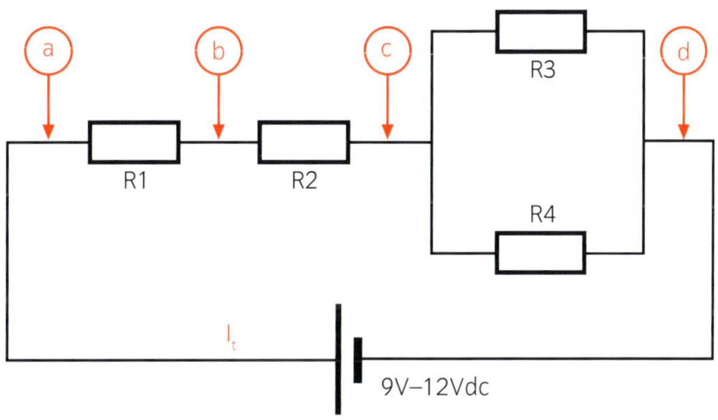

TOTAL CIRCUIT RESISTANCE:

Using the expression given below, calculate the total circuit resistance (R_t):

$$R_t = R1 + R2 + \left[\frac{R3 \times R4}{R3 + R4}\right] = \_\_\_\_\_\_\_\_\_\_\_\_ \Omega$$

Now prove this by measurement:

1 Do not connect the dc supply to the circuit.

2 Using a multimeter set to a suitable resistance range, measure the resistance of R1 + R2 by connecting the test probes to test points (a) and (c). Enter the value in the space below.

3 R1 + R2 = ____________ Ω

4 Using a multimeter set to a suitable resistance range, measure the resistance of R3//R4 by connecting the test probes to test points (c) and (d). Enter the value in the space below.

5 R3//R4 = ____________ Ω

The total measured circuit resistance can be found by adding measurement 3 to measurement 5.

Therefore R_t = ____________ Ω

Continued

E42 (CONTINUED)

RESISTIVE CIRCUIT THEORY

From your tests we can see the following:

- The measured value of R1 and R2 is equal to the added value of these two resistors.

Remember: for resistors connected in series, the total value is the addition of all of the resistors.

- The measured value of R3 connected in parallel with R4 is less than the lowest value resistor. If R3 and R4 have equal values, then the total resistance is half the resistor value.

Remember: for resistors connected in parallel, the current divides between the resistors, therefore reducing the circuit resistance.

Was the measured value of R_t exactly the same as the calculated value? Probably not. What are the possible reasons for the discrepancy?

____

____

____

VOLTAGE DROP (POTENTIAL DIFFERENCE) IN A CIRCUIT:

1. Connect a dc supply of between 9V and 12V to the red and black leads on your circuit.
2. Using a multimeter set to a suitable voltage range, measure the volt drop across R1 by connecting the test probes to test points (a) and (b). Enter the value in the space below.
3. V_{R1} = ________________ V
4. Using a multimeter set to a suitable voltage range, measure the volt drop across R2 by connecting the test probes to test points (b) and (c). Enter the value in the space below.
5. V_{R2} = ________________ V
6. Using a multimeter set to a suitable voltage range, measure the volt drop across R3//R4 by connecting the test probes to test points (c) and (d). Enter the value in the space below.
7. $V_{R3//R4}$ = ________________ V

   Remember: the sum of the potential differences in a circuit is equal to the applied EMF (E). Prove this by adding $V_{R1} + V_{R2} + V_{R3//R4}$ = ________________ V

Continued

TOTAL CIRCUIT CURRENT:

Using the expression given below, calculate the total circuit current (I_t):

$I_t = E \div R_t =$ ______________ mA

To measure a current in a low voltage dc circuit, you must break the circuit at some point and connect a multimeter set to a suitable current range across the break in the circuit. In other words, the multimeter is connected in series with the circuit current flow.

For the circuit in this task, disconnect the positive lead from the dc supply and connect your multimeter between the red lead and the power supply terminal (see figure).

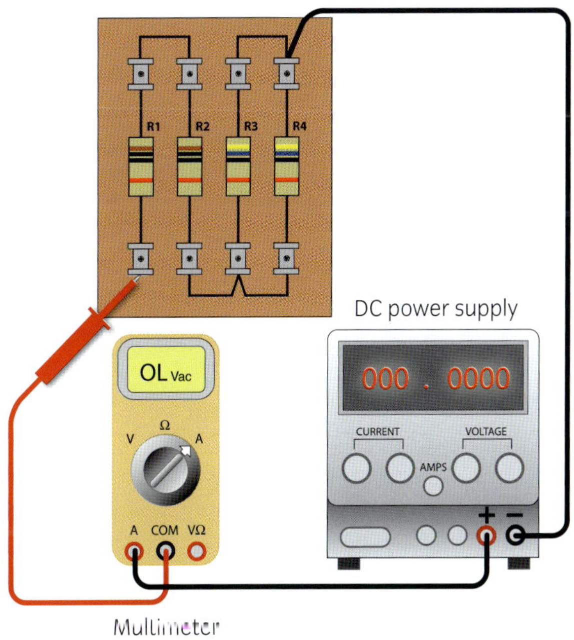

Measured value of $I_t =$ ______________ mA

How does the measured value of I_t compare with the calculated value? Comment on any discrepancies.

____

____

Date completed
Tutor/trainer's feedback/comments

E43

APPLIED RESISTIVE CIRCUIT THEORY

INVESTIGATE R_1 AND R_2 RESISTANCE BEHAVIOUR IN A RING FINAL CIRCUIT:

In this task you will construct the equivalent circuit for the line (R_1) and cpc (R_2) conductors in a ring final circuit containing three socket outlets (we shall assume that the CSA for both conductors is the same). Then, using a low resistance ohmmeter, you will perform continuity tests both with the circuit connected correctly, and incorrectly with an extra conductor creating an unwanted loop.

The following diagrams illustrate the two circuits that you will be simulating. Each cable will be represented by a resistor.

Note: for simplicity, we will not include the neutral circuit in this task.

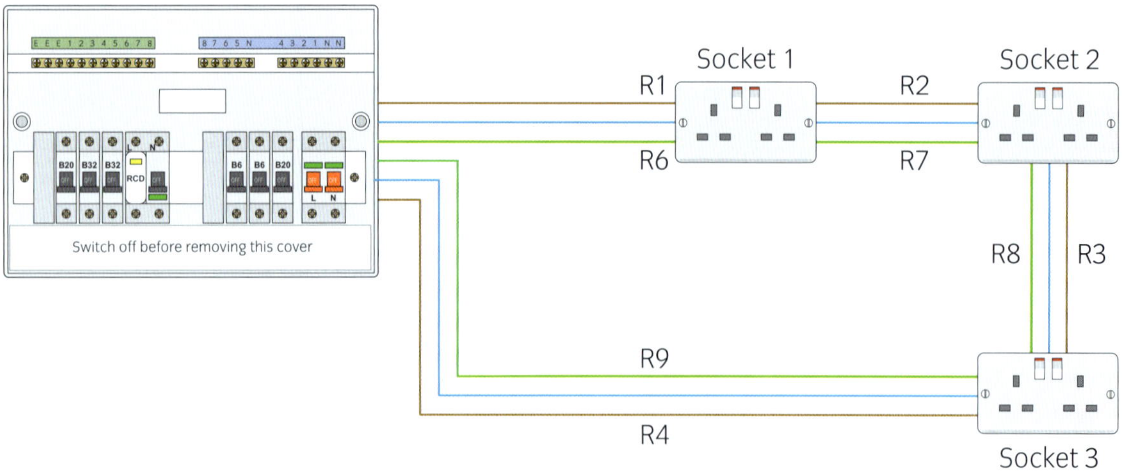

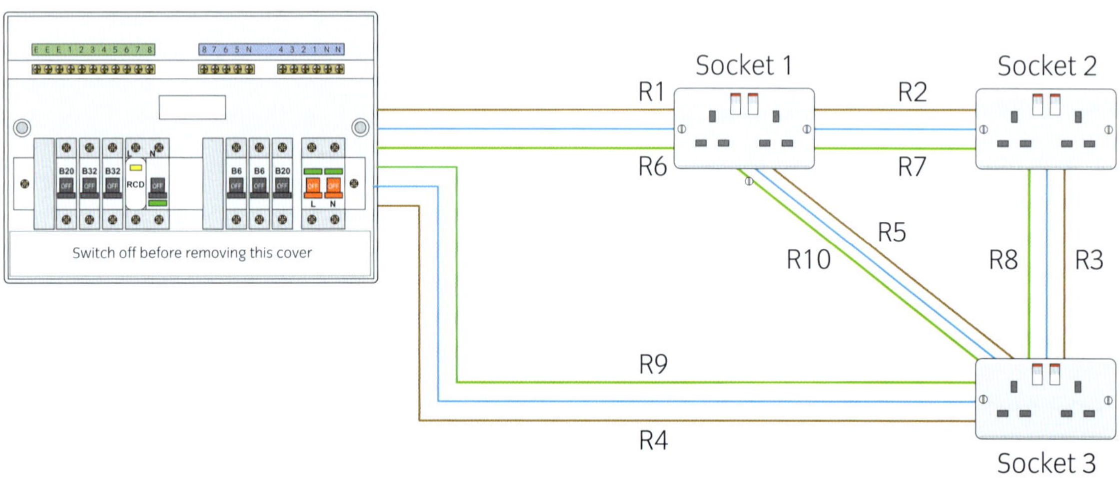

Continued

E43 (CONTINUED)

APPLIED RESISTIVE CIRCUIT THEORY

If we replace the line and cpc conductors with equivalent resistor values, we obtain the equivalent circuits shown.

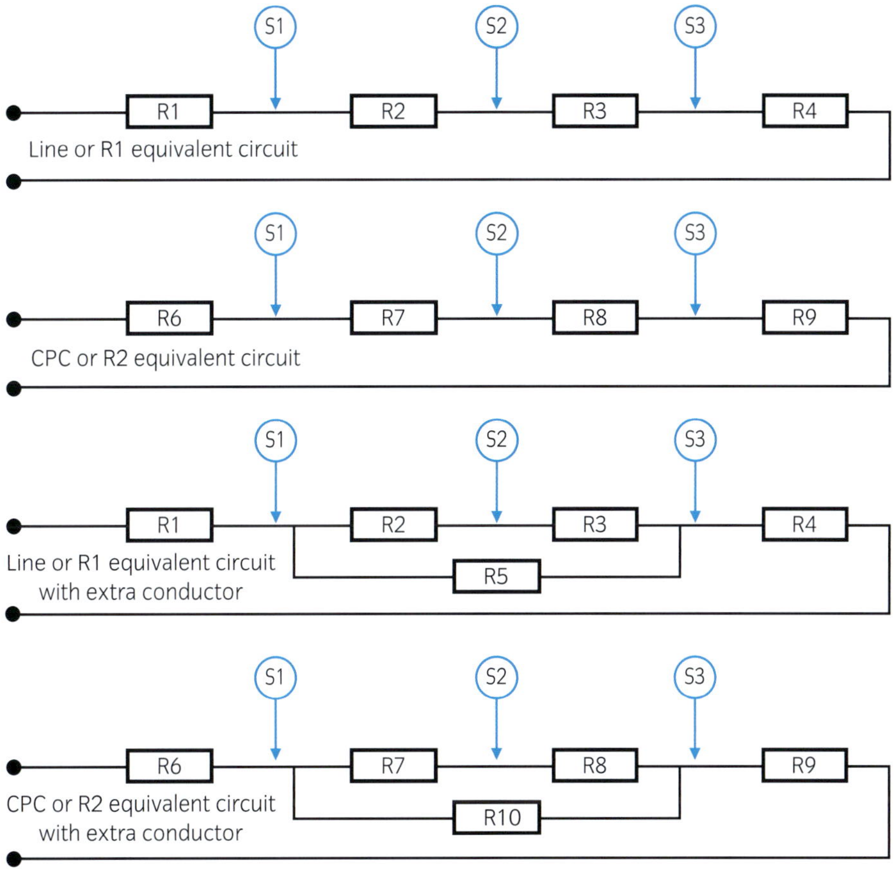

Continued

E43 (CONTINUED)

APPLIED RESISTIVE CIRCUIT THEORY

CONSTRUCT YOUR CIRCUIT:

An ideal resistor value to use for each conductor in our test circuit is 2.2Ω. However, your tutor/trainer may issue you with a slightly different value, in which case your readings will differ from the examples given in this task.

Item	Complete all the tasks below	Candidate check ✓
1	Mount resistors R1 to R10 onto the tag board in the positions shown in the layout diagram, wrapping each wire once around the tag using long-nosed pliers (the method for doing this is illustrated in task E40). Cut off excess wire ends. **Do not** solder any terminals at this stage.	☐
2	Use uninsulated wire in the order of 22 SWG to make links 1–8. Using long-nosed pliers, form each link to have square corners and suitable width to fit onto the tag board. Fix each link onto the tag board in the positions shown in the layout diagram, soldering each terminal only once all wires have been wrapped around it.	☐
3	Cut 250mm lengths of 22 SWG insulated wire: • 2 x brown for the line circuits • 2 x green for the cpc (you will not be able to obtain green/yellow in such a small gauge) • 1 x red for jumper link to create the 'extra conductor' link • 1 x black for jumper link to create the other 'extra conductor' link.	☐
4	Strip 10mm of insulation from both ends of each of these wires, using a suitable stripping tool.	☐
5	Tin all of these wire ends using a soldering iron (the method for doing this is illustrated in task E40).	☐
6	Fit one end of each of these wires to the terminals shown in the layout diagram by wrapping the end of the wire around the tag. Make sure that you fit each colour wire to the correct tab. Solder each wire as you fit it.	☐
7	Fit a crocodile clip to the end of each of the red jumper links.	☐

Continued

E43 (CONTINUED)

APPLIED RESISTIVE CIRCUIT THEORY

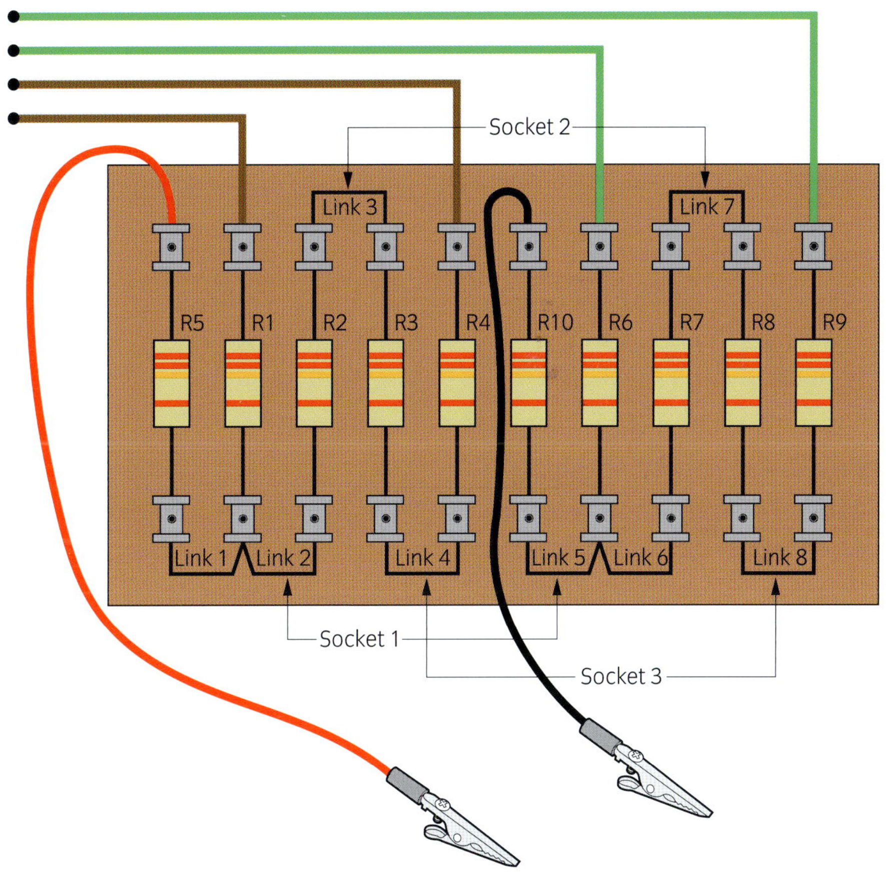

CONTINUITY TESTING:

The first test that you would perform on a ring final circuit would be the continuity of each loop (line, neutral and cpc). To do this you would remove all conductors and connect an ohmmeter across the ends of each conductor in turn, noting the loop resistance.

If you have used 2.2Ω resistors in your mock-up, then:

- for the correct circuit arrangement, the resistance of each loop will be 4 x 2.2 = **8.8 Ω**.
- for the incorrect circuit arrangement, the resistance of each loop will be:

$$R1 + R4 + \left[\frac{(R2 + R3) \times R5}{R2 + R3 + R5}\right] = 4.4 + \left[\frac{9.7}{6.6}\right] = 4.4 + 1.5 = \mathbf{5.9\ \Omega}$$

Continued

E43 (CONTINUED)

APPLIED RESISTIVE CIRCUIT THEORY

Prove these values by testing:

Item	Complete all the tasks below	Candidate check ✓
1	Without the jumper link crocodile clips connected (ie, for a correct ring final circuit arrangement), connect a low-resistance ohmmeter across the two brown conductors. This is your R_1 value for the ring final circuit. Enter the reading in the space below.	☐
2	Connect a low resistance ohmmeter across the two green conductors. This is your R_2 value for the ring final circuit. Enter the reading in the space below.	☐
3	Fit the red wire crocodile clip to link 4. This will add one unwanted conductor to your circuit.	☐
4	Fit the black wire crocodile clip to link 8. This will add the other unwanted conductor to your circuit.	☐
5	Connect a low-resistance ohmmeter across the two brown conductors to obtain a new R_1 reading. Enter the reading in the space below.	☐
6	Connect a low-resistance ohmmeter across the two green conductors to obtain a new R_2 reading. Enter the reading in the space below.	☐

Circuit condition	R1 loop resistance	R2 loop resistance
Correct		
Incorrect		

In a real-life situation, you would have estimated the value of resistance for R_1 and R_2 by estimating the length of the cables. If your measured values are close to the estimated values, then you can be sure that the circuit is connected correctly. In this example, your testing has proven to you that there is a fault with this ring final circuit.

Continued

E43 (CONTINUED)

APPLIED RESISTIVE CIRCUIT THEORY

PROVING CIRCUIT FUNCTIONALITY:

The second test that you would perform on a ring final circuit would be to cross-connect the line and cpc conductors at the distribution board, and measure the resistance between line and earth at each socket outlet.

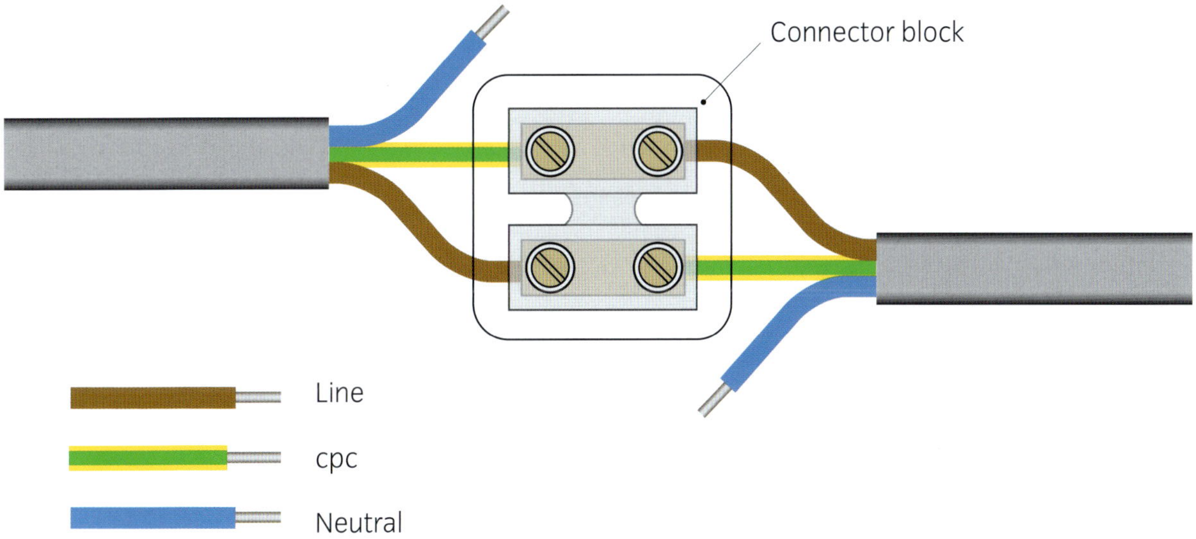

For a correctly connected circuit, the reading taken at each socket will be more or less the same, and will be ¼ that of $R_1 + R_2$. That is: 0.25 x (8.8 + 8.8) = **4.4 Ω**.

The following diagram below shows your ring final circuit with L and E cross-connected. The dotted resistors represent the circuit created by the extra conductor between socket outlets 1 and 3.

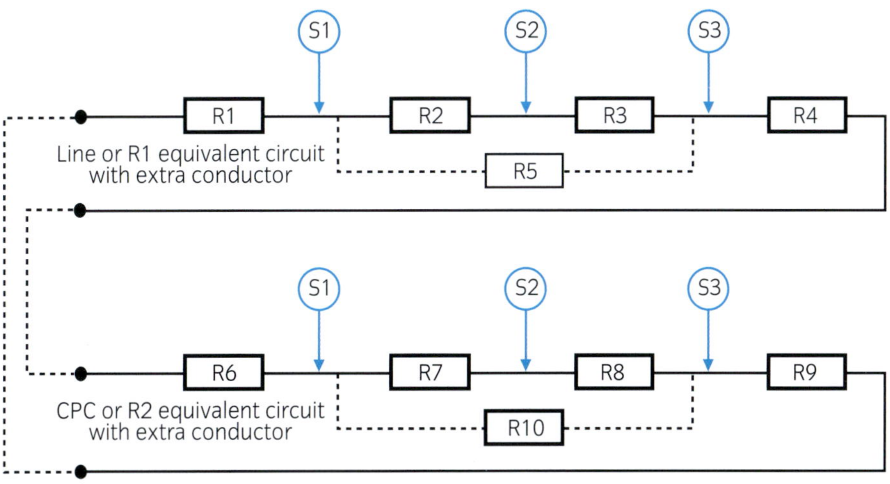

Continued

E43 (CONTINUED)

APPLIED RESISTIVE CIRCUIT THEORY

Item	Complete all the tasks below	Candidate check ✓
1	Without the jumper link crocodile clips connected (ie, for a correct ring final circuit arrangement), connect a low resistance ohmmeter between the two points labelled S1 (socket 1, L and E). Enter the reading in the space below.	☐
2	Connect a low-resistance ohmmeter between the two points labelled S2. Enter the reading in the space below.	☐
3	Connect a low-resistance ohmmeter between the two points labelled S3. Enter the reading in the space below.	☐
4	Fit the red wire crocodile clip to link 4. This will add one unwanted conductor to your circuit.	☐
5	Fit the black wire crocodile clip to link 8. This will add the other unwanted conductor to your circuit.	☐
6	Repeat the resistance tests in steps 1–3 above, entering your readings in the space below.	☐

Item	Correct circuit connection	Incorrect circuit connection
Socket 1 L-E		
Socket 2 L-E		
Socket 3 L-E		

Date completed

Tutor/trainer's feedback/comments

E44

THREE-PHASE SUPPLY MEASUREMENT

MEASURE VOLTAGE, CURRENT AND PHASE SEQUENCE OF A THREE-PHASE SUPPLY:

Note: these instructions provide an outline sequence of tests that you might perform when testing a three-phase, 400V electrical supply. However, the actual method that you use will be determined by the design of the test rig and the type of test instrument(s) that you are using in your centre.

Important: because of the high voltages and currents associated with this testing, this task must only be carried out under the supervision of your tutor/trainer.

The three tests are:

- voltage
- current
- phase sequence.

VOLTAGE TESTING:

Item	Complete all the tasks below	Candidate check ✓
1	Select the appropriate voltage test instrument and perform a visual check of the instrument and all test leads and connectors to ensure that it is safe to use.	☐
2	Set the instrument to the appropriate range.	☐
3	Complete Table 1 below by testing the voltage between each of the terminals listed in the table.	☐

Continued

E44 (CONTINUED)

THREE-PHASE SUPPLY MEASUREMENT

Table 1

Terminals	Measured voltage
L1–neutral	
L2–neutral	
L3–neutral	
L1–L2	
L1–L3	
L2–L3	

CURRENT TESTING:

Item	Complete all the tasks below	Candidate check ✓
1	Select the appropriate test instrument and perform a visual check of the instrument and all test leads and connectors to ensure that it is safe to use.	☐
2	Set the instrument to the appropriate range.	☐
3	Complete Table 2 below by testing the current in each of the conductors listed in the table.	☐

Continued

E44 (CONTINUED)

THREE-PHASE SUPPLY MEASUREMENT

Table 2

Conductor	Measured current
L1	
L2	
L3	
Neutral	

PHASE SEQUENCE TESTING:

Item	Complete all the tasks below	Candidate check ✓
1	Select the appropriate test instrument for this test and perform a visual check of the instrument and all test leads and connectors to ensure that it is safe to use.	☐
2	**Confirm with your tutor/trainer that the 400V supply has been isolated before proceeding**. Connect the test leads in the correct sequence to the three supply terminals.	☐
3	Set the instrument to the appropriate range.	☐
4	Energise the 400V supply and confirm that the phase rotation for this supply is correct.	☐
5	**Confirm with your tutor/trainer that the 400V supply has been isolated before proceeding**. Reverse the test leads connected to L1 and L3.	☐
6	Energise the 400V supply and note the phase rotation indicated on the test instrument.	☐
7	Confirm that the 400V supply has been isolated before removing the test leads.	☐

Date completed

Tutor/trainer's feedback/comments

E45

COMPONENT IDENTIFICATION

This task could be performed in pairs or small groups.

Identify electrical and electronic components.

Your tutor/trainer will provide you with ten components from the list below. The components will be laid out on a surface with the numbers 1 to 10 against them.

Enter the number of each component next to its name in the list, and write a brief description of at least one common application for that component. If you are uncertain of a component, or its application, use available resources such as your course notes, City & Guilds SmartScreen, or the internet to find the answer.

Number	Component type	Typical application
	Wire-wound resistor	
	Carbon resistor	
	Variable resistor	
	Polyester capacitor	
	Electrolytic capacitor	
	Light emitting diode	
	Diode	
	Inductor	
	Transformer	
	Relay	
	Solenoid	
	Switch	
	Fluorescent starter	
	Filament lamp	
	Thyristor	
	5 Amp terminal block	
	30 Amp terminal block	
	RCD	
	MCB	
	RCBO	

Date completed

Tutor/trainer's feedback/comments